KELY
KNOWLES

Twelve Men Down:
Massachusetts Sea Rescues

by Robert H. Farson

Cape Cod Historical Publications
Box 281
Yarmouth Port, Massachusetts, 02675

Library of Congress Catalog Card Number: 99-85833

Designed by Joan Hollister Farson

Frontispiece: Highlands Life-Saving Station, Cape Cod
Painting by Kely Knowles

Printed in the United States of America

ISBN #0-9616740-2-4

First Edition

Cape Cod Historical Publications
Box 281
Yarmouth Port, Massachusetts
02675

Dedication

To my Mother, Annie Cheves Farson. 1881-1943.

She was born and lived much of her life by the sea. Her dedication to her family was total.

To my wife, Joan Hollister Farson.

Her dedication to this book was a wonder to see.

TABLE OF CONTENTS

INTRODUCTION

When you look at the past it seems if there was a problem it had to continue for a long time, often with serious consequences, before anything was done about it. Such was the case of shipwrecks. Like so many other problems, the Chinese were the first to attempt a solution. Along some of its rivers a system of lifesaving was begun years before it started in Europe. In 1708 an association was formed up the Yangztse River above Shanghai that created the first lifesaving system that had stations, lifesavers and rescue craft. Private benevolent groups paid for these organizations and the Chinese government set up lifesaving paid for by land taxes. So in China the two types of organizations involved in lifesaving were created: private and state-sponsored.

In Europe in the 1760s the Dutch started limited lifesaving followed by the British some eight years later. It wasn't until 1824 the British used lifeboats in their Royal National Lifeboat Institution. The interest in such work grew and a number of pioneer efforts in lifeboat design came from it.

The idea of lifesaving spread to Boston after the American Revolution where the Massachusetts Humane Society was formed as a voluntary group. The State of Massachusetts occasionally gave some financial support, but the Society subsisted on private funds, mostly from its members. Boat houses and shelters for the shipwrecked victims were built along the coast; volunteers were trained to respond.

Despite the horrors of some shipwrecks, the Federal Government did not become involved until it almost had no choice. Public attitudes left it no way to continue ignoring the problem. Finally after limited construction of stations along the New Jersey shore but no provision for staff, the United States Government provided money to build a professional lifesaving service in the early 1870s. The system spread to all coasts and the Great Lakes. The media of the day, newspapers and magazines, became interested in the United States Life-Saving Service. They spread the word about rescues under some of the worst conditions. When surfmen perished in their work, the American public responded with money to help their families.

America's greatest artist of the Nineteenth Century, Winslow Homer, painted several canvasses of lifesaving, the most famous of which is The Lifeline.

After more than forty years the Life-Saving Service was moved into the newly created Coast Guard in 1915. The principal focus of this book is the Humane Society and the U.S. Life-Saving Service along the coast of Massachusetts from Cuttyhunk and Martha's Vineyard to Salisbury Beach, the northernmost station in the Bay State.

Robert H. Farson

ACKNOWLEDGEMENTS

Joan H. Farson, Richard Boonisar, Cape Cod National Seashore, Philip Choate, James Claflin, Admont Clark, Daniel Davidson, John Perry Fish, Harold Flagg, Maurice Gibbs, R. Loren Graham, Barry Homer, William H. Hough, Hull Life-Saving Museum, Kely Knowles, Linda McCausland, Hope Morrill, Paul C. Morris, Lori Newmyer, Joseph Nickerson, Charlotte Price, William P. Quinn, Mary Sicchio, Bonnie Snow, Edward Robinson, Ted Rose, Don Spring, Gretchen Widegren and the Yarmouth Port Library.

THE ILLUSTRATIONS

The illustrations, collected over many years, are from a variety of sources including Bath, Maine Marine Museum, Richard Boonisar, Bourne Historical Society, Cape Cod National Seashore, Chatham Historical Society, Alexander Cheves, Vera Cheves, Philip Choate, Annie Cheves Farson, Harold Flagg, R. Loren Graham, William H. Hough, Hull Life-Saving Museum, Paul C. Morris, Nantucket Life-Saving Museum, Nickerson Room of Cape Cod Community College, Orleans Historical Society, William P. Quinn, Don Spring, United States Coast Guard, and the United States Life-Saving Service. The collection of the author, who also drew the maps.

MASSACHUSETTS AND THE SEA

The migration from Europe to American colonies was along the coast. All of those who came arrived by sea and they lived by the sea. New England soil was difficult to cultivate, and the ocean offered opportunities. During the 1600s the great availability of fish brought Europeans to Massachusetts in large numbers who wound up sailing to the largest fishing grounds, the Grand Banks, off Newfoundland, an area of 40,000 square miles.

Next was Georges Bank east of Cape Cod with 10,000 square miles. The settlers had to look to the sea for their survival because they had no major crops like sugar in the West Indies or tobacco in Virginia.

To go to sea the settlers needed ships and that meant timber, which New England had in vast quantities. So shipbuilding became the foundation of the maritime life. It was concentrated at the mouths of rivers. Before 1720 sloops and ketches were the most popular types. After that schooners became the largest number of two-masted vessels produced. By the time of the American Revolution, New England built two-thirds of the ships produced in the colonies.

The earliest explorers had favorite parts of the coast with John Smith writing about the shore of Massachusetts Bay "as a paradise." Bartholomew Gosnold was partial to Buzzards Bay, especially Cuttyhunk.

During the great age of sail from the 1600s to the late 1890s the sailor's life was the most dangerous calling a man could choose. Maritime men studied the weather because around New England conditions change more often through a greater range of extremes with bad weather tending to be worse north of Cape Cod. The problem with New England coastal weather is the converging of two storm systems over the area: one from the west around the Great Lakes heading east, and the other pushing up the Atlantic seaboard from the south. When the two meet, as happens, a storm of incredibly violent proportions can churn the ocean into huge waves and titanic violence.

That has happened twice since near the end of the nineteenth century when the hurricane known as the *Portland* Storm occurred off Massachusetts from Vineyard Haven to southern Maine the weekend after Thanksgiving, 1898. The other event was even more violent when the same conditions created the Halloween Storm of 1991, now immortalized by author Sebastian Junger in his book, “The Perfect Storm.” Both of these events showed the dark side of nature.

When the wind shifts to the east the waters are whipped into a nasty chop and then high swells roll in from the ocean. In the winter, gales are more frequent than other seasons. Conditions then are worse not only because of the cold but the wind-driven wet snow reduces visibility to near zero at sea.

The summers have their own problem: it is fog. The ocean from Nantucket to northern Maine breeds fog when warm moist air is carried north by southerly winds where it meets cold air chilled by waters from the Labrador current. Around Chatham and Monomoy there is heavy fog for more than 160 days a year. The lightships' foghorns around Pollock Rip and Handkerchief Shoals used to operate in one of every two hours during July. In such conditions the fogbound sailor lives in a private world limited to fifty feet away. He could not see beyond that.

Thousands of hapless vessels were destroyed during violent gales, striking sunken ledges or stranding on sandy shoals or lost in a heavy fog. The busy sailing port of Salem, Massachusetts had more than four-hundred widows. There were many women without husbands and children without fathers in Marblehead, Gloucester, Newburyport, New Bedford, Nantucket and Hyannis-where so many men did not return from a voyage.

Maritime New England's major industry was fishing, and the earliest centers of the business were Scituate, Nantasket and Barnstable. By 1765 the cod fishing industry, centered in Massachusetts, sailed five-hundred ships, employing five-thousand men. Each year the total catch was worth $1-million. Marblehead was the leading fishing port with Gloucester next. Those two towns accounted for sixty percent of the catch. There were twenty other towns that sent out cod boats.

Whaling began its growth with close to shore catches at first with the whales brought ashore and grounded on beaches with the blubber cut up and boiled for oil, which was used for illumination. This began to take shape in the late 1600s. By 1700 whaling began to move offshore. It was almost totally a Massachusetts affair with Nantucket and Dartmouth, later New Bedford, dominating the business and accounting for eighty percent of the whale oil shipped to English markets. That was not surprising, as Massachusetts dominated New England Colonial commerce. For one example the prosperity of Newburyport increased fourfold between 1793 and 1807. That was the town's total valuation.

The earliest effort to reduce the danger of coastal sailing was the building of lighthouses, with the first one at the entrance to Boston Harbor on Little Brewster Island. That was in 1716. Other places followed: on Brant Point, Nantucket in 1746 and later Gurnet near Plymouth. Vessels using the harbor contributed to the upkeep of the light and to the salary of the keeper.

Those were built before the Revolution. In 1789 those lighthouses were turned over to the new federal government. Local maritime people placed buoys and day beacons marking channels. Private people placed buoys marking channels around Salem, Massachusetts. The state of Massachusetts placed buoys in the Merrimack River, leading up to Newburyport and beyond. The state built a lighthouse at Great Point, Nantucket, while in 1789 twin lights were built on Thatcher's Island off Rockport.

In spite of these aids to mariners, accidents to ships continued at an increasing pace with vessels breaking up on sunken ledges and shoals. From Plum Island to Nantucket Sound the toll of lost ships increased at an alarming rate. Conditions had to be right for survival of the crews. During the summer it was an easier time, especially if the accident was near the shore.

The winter was a different matter. Today it is called hypothermia. In the eighteenth century they knew it as freezing to death in bitterly cold water of the Massachusetts coast. On those sailing vessels the crew climbed the rigging and held on above the waves crashing over the hull below, until, unseen from land, one by one they froze to death in the high winds. Many sailors could not swim and the small boats that vessels carried were not designed to save lives in wild water breaking around their stricken ship. To try to reach shore that way was almost certain disaster. In fact, when finally a professional life-saving organization was formed much later, their advice to sailors was to stay with your ship and we will come to your rescue in surfboats designed for such emergencies.

In Massachusetts and elsewhere marine societies were active in every port to aid widows and other survivors whose relatives died at sea. Ship captains were asked to make charts of areas they had sailed and to keep logbooks for the use of others and recommend locations for lighthouses and then lobby for their construction. By 1800 merchants and ship owners had formed marine insurance companies, which shared risks among many policy holders.

The port of Boston was the largest in New England. All around the miles of waterfront were docks, warehouses, grain elevators, coal pockets, shipways, sail makers, ships' chandleries, and companies manufacturing masts and spars, rope, chains and anchors, signal lanterns, pumps and ship fittings, water casks, boilers, hoisting engines, wire for standing rigging, cabin stoves, sextants and chronometers and dealers in ship timber like yellow and white pine, oak, spruce, hackmatack, teak, mahogany, rose wood, black walnut, ash, and birds eye maple.

Boston was indeed a maritime city where the many newspapers had pages of advertisements and information of arrivals and departures of ships with cargoes and destinations or origins, contracts for vessels and word of new businesses, aids to navigation, dredging and all the vast information that a maritime center needed every day.

Boston was home port to a fleet of deepwater vessels and to the growing coastal fleet of small fore-and-aft rigged sailing ships that were economical to run and easily built from local timber. They were short-haul freight and passenger vessels and their contribution to coastal life was considerable.

Close to shore the many vessels of the coastal fleet sailed day and night into a bright morning and into the darkest gloom of past midnight with the tide and then felt the wind increase and saw the dark clouds of a storm rising. Into all kinds of New England weather they sailed. For as long as men sailed the oceans they came to fear and respect it. The fear they held was earned.

We do not know with any precision the toll in ships and sailors and passengers lost at sea. So many vessels described as "gone missing,"-simply vanished, leaving no traces, just questions.

In May 1784 a visitor arrived from London who brought with him a suggestion for Boston's maritime community. The man was Dr. Moyes and he suggested the outline of a plan to form a society similar to The British Royal Humane Society, incorporated twelve years before. The British Society was an imitation of one formed before that in Holland "To restore to life such persons as were apparently dead from drowning, or any other sudden means of the extinction of life." The Dutch group was formed in 1767 in Amsterdam. Dr. Moyes detailed many cases of complete recovery in Great Britain.

The English physician's idea found sympathetic listeners and on January 5, 1786 the first meeting of the subscribers was held at the Bunch of Grapes Tavern in State Street,

Boston. This favorite local hostelry was at the corner of King Street and Mackerel Lane.

They agreed to no more than twelve trustees of whom one was James Bowdoin, the governor of Massachusetts. The others nominated and then elected Bowdoin the first president of the Humane Society of The Commonwealth of Massachusetts. Bowdoin was esteemed equally as a philosopher and statesman. He was a fellow of The Royal Society of London, an early friend of Benjamin Franklin and correspondent of George Washington, with whom he discussed the possibilities of a Cape Cod Canal. Bowdoin represented and helped to fix all the best traditions of the community. The second name on the earliest list of officers was that of Dr. John Warren as second vice-president is one which may stand beside Governor Bowdoin. Warren was a brother of the hero of Bunker Hill. He was a founder of the Harvard Medical School in 1783. Warren was the first Professor of Anatomy and Surgery. One of his sons later said,"The Humane Society was one of my father's strongest objects of interest."

From 1786 to 1916 the Society awarded 179 gold medals, 1029 silver medals, twelve silver plates, 511 certificates, 705 bronze medals. In addition the cash awards to those who had helped rescue others over the years 1787 to 1916 totaled $28,229.50.

Along the Massachusetts coast the Merrimack Humane Society of Newburyport was founded in 1802. Its work was similar.

The Massachusetts Humane Society built small, wooden houses of refuge along the coast for survivors of shipwrecks. It was the first project of the Society.

Elsewhere in the New World a humane society was formed in Philadelphia in 1780, a group that posted signs along the waterfront of the Delaware River telling where its lifesaving apparatus could be found. In hot weather the Philadelphia group posted signs near taverns, warning of the dangers of cold water. In 1794 similar societies were founded in New York City and Halifax, Nova Scotia.

The Massachusetts Humane Society became the model for lifesaving by shortly after it was formed building small houses of refuge near the shore to shelter shipwrecked sailors. The first of these was constructed in 1787 on Lovell's Island in Boston Harbor with several more at Nantasket and Scituate. By 1790 three more were added. Two years later another was built on Cape Cod.

In 1794 another house of refuge was built on Nantucket's Great Point about a mile and a half south of the lighthouse at the end of Sandy Point. Nantucket received a second such building on the south shore. These one-room sheds had a fireplace with a chimney and were stocked with wood for a fire, sperm oil and a lamp, casks of water, hard tack and blankets.

The Reverend Thomas Barnard, minister of the North Church of Salem, in 1794 gave a discourse concerning the Society's work:

"... the Society has extended its kind providence to the shipwrecked sailor. Our brethren of the sea have peculiar claims upon us, for our humane attention. This coast is eminently dangerous at the cold and stormy season of the year. And then our sailors usually meet with their most cruel disasters Yet at this season, they must be upon it, as well as the more safe and pleasant. With courage, they expose themselves to all the risks and dangers of a most adventurous situation. By their labors, they are a rich source of convenience and wealth to their country; and by their resolution and hardy habits, a powerful means of defending it, when attacked. Say ye, who feel for another's woe! Is not the shipwrecked sailor an object deserving your tenderest pity, your generous compassion! The recovery of land which he beholds with eager eyes, appears only the means of safety...he plunges into more threatening danger, that he may seize it. He succeeds; he treads again upon firm ground; and for a moment thinks himself safe. But wet, cold and hungry...is convinced he must die...unless he is relieved. At this moment of high distress, with what transport of joy must he behold the humble shed, which we have erected; and find a supply of the necessaries he needs. Then, the blessing of him that is ready to perish, will descend upon us."

A persistent problem with the refuge huts and for the Society was the robbing of the materials placed inside them for the victims of shipwrecks. As early as 1790 the Society appealed to the public to "protect them from lawless depredations; and whereas said Houses are frequently robbed of the means necessary for the relief aforesaid by Persons destitute of the feelings of humanity, the Trustees request those who reside in the vicinity of the said Huts to use their utmost care that the articles deposited therein may remain uninjured." Those feelings were published in Boston by order of the Trustees and signed by the Recording Secr'y John Avery, Jun.

In 1802, because of the frequency of wrecks on Cape Cod's Great Beach, five more houses of refuge were built from Race Point to Monomoy. There were poles and signs erected to show the way to the shelter for those reaching the beach.

But the Humane Society had greater ambitions than just providing small sheds or shelter. They decided to have a lifeboat built and a building to house it. A builder of whale-boats on Nantucket, Henry Raymond, under the supervision of Captain Gideon Gardner constructed the first American small craft ever built especially for lifesaving. It was copper-fastened, thirty feet long with a ten-foot beam with ten shortened oars and steering oars. She was rowed by ten men, double backed. The craft cost $ 1,433.11. It was placed at the Humane House in Cohasset in 1807. The shed to protect it cost the Society $160.

The Society continued to build the Humane House shelters along the Massachusetts coast with a total of thirteen on Nantucket alone. Notices were posted in seaports and nautical pages in newspapers calling attention to their locations.

Later the Humane Society built boathouses with surfboats and breeches buoys with Hunt Guns on beach apparatus at the right.

With a greater need to assist in the rescue of sailors in peril the Humane Society placed lifeboat stations from Plum Island to Cuttyhunk, including Martha's Vineyard and Nantucket. In the next seventy years after the Cohasset Station was established more than eighty stations were built and manned by volunteers from among local citizens, fishermen mostly and others who were good watermen with knowledge of small boat handling. During that period from 1840 into the 1870s the construction of lifeboat stations was the major goal of the Society, an era referred to in retrospect as "the golden age of the Massachusetts Humane Society." It was a vigorous effort with the appointment of responsible men as keepers, often with one keeper responsible for several boathouses. A number of these were built adjacent to lighthouses as at Annisquam, Plum Island, Eastern Point, at Gloucester, the western end of Cuttyhunk and across Vineyard Sound at Gay Head. Many of these men did double duty. The lighthouse keeper was often the Humane Society keeper. Cash awards were made to those involved in rescues with a sliding scale according to the difficulty of the operation. Medals and certificates were awarded as well. Teams from Cuttyhunk, Nantucket and Hull won more decorations than the others. Their locations were more perilous than others.

From time to time older boats were replaced with newer models of the surfboat size-under 25 feet-and for some of the stations dories and mortars for line throwing were provided as well.

With the formation of the United States Life-Saving Service in the early 1870s the growth of the Society's number of lifeboat stations slowed to a stop, but the volunteer groups remained in place and often joined the new professionals of government service when extra hands were needed at difficult rescues. At some shipwrecks Society volunteers were the only ones to show up. Which is what happened when the big schooner *Nancy* stranded on Nantasket Beach in February, 1927. That rescue is told in a separate chapter.

As the need for nautical lifesaving dwindled the volunteer groups gradually drifted away in the 1930s, and by the end of the decade the Massachusetts Humane Society's ocean lifesaving efforts were discontinued. Its work was a model for government operations and even today its pioneer efforts along the 192 miles of the state's coastline brings the highest praise from many sources. Books and articles continue to be published and always the work of lifesaving by the Society is hailed as the first shore-based rescue effort in America some seventy years before the United States Government reluctantly began an effective service. As Dr. Robert Haley writes:"The legacy of this unique society still has a profound influence on many institutions in the Commonwealth, and its actions can truly be said to have made the Boston area the original home of organized lifesaving in the United States."

Surfboats were designed for local conditions from Cuttyhunk to Plum Island. Most rescues used them as Massachusetts was surfboat country.

THE HUMANE SOCIETY AND THE U.S. LIFE-SAVING SERVICE

Sailors have always faced danger. Along the East Coast with few lighthouses and other aids to navigation ship captains navigated by coasting which means they sailed close to the shore to find landmarks to fix their position. Sometimes they were caught with a strong onshore wind and were driven ashore or at least aground on a bar. In any case the vessel was usually lost. That was coasting and it had its perils.

The biggest port on the East Coast was New York and the approaches to it had two dangerous coasts, New Jersey and Long Island, each had sand bars. Off New Jersey, those were anywhere from one-thousand feet to twenty-five hundred feet from the beach. A sailing vessel grounding that far out broke up quickly in the heavy surf. If anyone could survive in the water that distance from shore when they reached the beach there was rarely anyone to help them, few people lived there along the shores.

In heavy surf close to shore the best rescue boats were small, especially designed, wooden craft, powered by men with oars.

A New Jersey congressman and medical doctor, William A. Newell, asked the House Commerce Committee to look into the possibility the federal government might help those in peril from shipwrecks. Newell's district ran from Sandy Hook to about half way down the coast, a shore that witnessed many accidents and deaths. Dr. Newell, so the story goes, was on a call near the shore when a sailing ship grounded on one of those off-shore bars too far for rescuers to reach. All on board died. His resolution before the committee failed to pass in 1847. The following year he introduced an amendment to a lighthouse bill which the House passed, calling for a ten-thousand dollar allotment for surfboats and other equipment to protect life and property along the shore of Newell's district in New Jersey. The money also paid for the building of eight lifesaving stations of modest size in that area. The United States Revenue Marine Service supervised their construction.

This move finally brought the federal government for the first time into the lifesaving business. The great impetus for passage of this bill was the combined efforts of the marine insurance industry in New York City and the commercial maritime interests. Their losses were considerable and escalating. These men represented money and power and were not to be ignored by the central government.

Now there was real movement as seven months later a further appropriation of twenty-thousand dollars passed to build and equip an additional six stations for the southern end of the New Jersey coast on the Atlantic Ocean.

A year later Congress passed a bill to include new stations on the Rhode Island shore at Watch Hill; North and South Carolina, Florida, Georgia and Texas. The Watch Hill station was New England's first. There would be more.

Later lifesaving gear and stations were approved for the Great Lakes, but there was no money for staff, nothing for keepers to be in charge or surfmen to actually rescue people.

Buildings and equipment, that's all. It wasn't enough. What the government provided was then turned over to local communities. They would raise volunteer crews to respond to their emergencies. It did not work. The stations fell into disrepair through neglect, equipment was stolen or misused and when there was a shipwreck no one was responsible and the heavy loss of life and property continued. This went on for four years. Actually longer than that.

In 1854 there were two maritime disasters with heavy loss of life. In the first one more than two-hundred emigrants died off the New Jersey coast; later that year two-hundred and thirty people perished when their vessel broke up in a winter storm. The Congress responded with a bill that provided keepers for the stations. Their pay was to be two-hundred dollars a year. However, since there was no appropriation of funds until three years later, there were no keepers until three years later.

In 1857 an appropriation provided for the salaries and local men were hired, but there was still a weakness: no surfmen were provided for. The new system worked like this: if there was a ship ashore, the keeper's job was to round up enough volunteers- at least six, eight was better- and launch the boat and head for the wreck. This was the theory, but it did not work well. The keeper had to walk a great distance to find anyone. This took too much time. Rounding up even a minimum number of men to help could take hours at a time when a ship was breaking up in a storm and everyone on board lost.

The coming of the Civil War had the attention of the nation in the late 1850s and the lifesaving program, such as it was, languished, until a year after Appomattox when William Newell, still a New Jersey congressman, convinced the lawmakers to approve a ten-thousand dollar appropriation to build additional lifesaving stations along the coast of his district.

In 1870 a bill to hire surfmen passed which provided for a half dozen lifesavers to be on duty at alternate stations for three months a year, beginning the middle of December. Their pay was forty dollars a month. However, the stations continued to go to hell through general neglect, even though each station had a keeper, many of them lived too far from the station and often were political appointees with little or no skill in small boats.

The dawn of a new era was not far off. In the winter of 1870-71 violent weather along the East Coast and the Great Lakes created havoc for the shipping industry. Vessels and their crews were destroyed in large numbers and finally the government decided to act, belatedly. It may have been a response to strong newspaper editorials calling for action to revamp the service so mariners in distress could depend upon help in the future. The marine insurance underwriters and the shipping people in New York continued to apply pressure through lobbying and Congress finally abandoned its plan for volunteers and passed a two-hundred thousand dollar appropriation to hire experienced surfmen along the shores of New Jersey and Long Island.

Let's go back for a moment to 1869 when the Treasury Department created a Division of Revenue Marine which included the revenue cutters-vessels to guard the coast as originally put to sea by President George Washington; steamboat inspection, marine hospitals and the lifesaving stations. The Secretary of the Treasury, George S.Boutwell, needed a man to head up this new bureau and he began his search one year later.

Boutwell took his time, almost a year, and then he offered the job to a thirty-six year old lawyer from Maine, Sumner Increase Kimball. The Secretary's choice was a ten-year career clerk in the Treasury. The one thing that Kimball insisted was that he would not be second-guessed on decisions involving personnel and reforms. "You will require an uncommon display of backbone," Kimball said to the Secretary. "But if you back me, I will put the bureau where you want it and where it ought to be." Boutwell replied,"no matter what the political pressure may be, I shall not interfere."

Sumner Increase Kimball in the 1870s became the General Superintendent of the U. S. Life-Saving Service and was its leader until it merged into the Coast Guard in 1915.

With that agreed to, Sumner Increase Kimball accepted the position as General Superintendent of the Revenue Marine Bureau, which included the brand new coastal life-saving branch of the Treasury Department, an organization in sad shape and badly in need of reform. And reform it, Kimball did. Armed with a report from an inspection of each station, Kimball asked for major changes from the volunteer system to a federal organization with professional standards, training and personnel. Many of the keepers were political appointees with little knowledge of small boat handling. Some of them were too old for rescue work. A number of them lived too far from the stations, which displayed a disgraceful neglect of equipment, much of which had been stolen, and everything filthy.

In 1872 nine life-saving stations were authorized for Cape Cod and by the following year they were in service. Later four more were added on Nantucket, and one each on Martha's Vineyard and Cuttyhunk. Others were added from Monomoy Point all the way to Salisbury Beach, near the New Hampshire line. With the only floating station on the east coast at Dorchester Bay, Boston; Massachusetts had a total of thirty-two life-saving stations in the federal rescue service.

Keepers were hired for twelve months and surfmen, usually six of them, for ten months of the active storm season that was changed and wound up being August first to May thirty-first. The keeper was a manager, a bookkeeper, a trainer of personnel, an expert in the management of surfboats and all the other equipment at the station, a keeper of all the activities, including wreck reports, of the station, he had to select the surfmen and oversee their professional activities. Most of the keepers were older men-at least thirty-and had to be no younger than twenty-one or older than fifty-five.

Their pay reached a high of nine-hundred dollars a year in the late 1880s and was never raised until the service was absorbed into the newly-created Coast Guard in 1915. Surfmen were paid fifty dollars a month for the ten months of the active season and that pay was raised to sixty dollars fairly early and never raised again. They could be paid three dollars for every rescue they worked if called in during the inactive period of June and July.

Keepers appointed the surfmen who were given numbers, one being the most experienced with number six the newest man with the least experience at the station. The seventh lifesaver was "the winter man" who served from December First through April. He usually stayed at the station to watch things after helping with the launch and later on their return to help beach the surfboat and aid anyone rescued.

The coast was divided into districts. Massachusetts was District two and had a District Superintendent, who for many years was Benjamin C. Sparrow of East Orleans. Sparrow selected the keepers and visited each station at least five times a year. He was not only a busy administrator but frequently was on hand during rescues. His annual pay was one-thousand dollars. When the government built its own telephone system about 1890, his home was on the circuit as were the halfway houses for the beach patrol and the lighthouses.

A word now about the beach patrol which was a major innovation of the Service. From August First to May thirty-first, the active storm season, surfmen walked the beach from dusk until dawn every day. The stations were usually placed no farther apart than five miles with two men leaving at once, one in each direction, north and south usually, and meeting half that distance with a surfmen from adjoining stations. Their meeting place was a halfway house, a small one-room, wooden building with a fireplace, a chimney and a supply of wood for warmth and a door facing away from the water and the wind. The surfmen carried a small brass check with their number on it. They exchanged these to prove they had made the walk. In some direction from a station the beach walk would be to the end of land and no one was coming toward the surfman. In those cases there often was a watch house with a key to record the time he was there on the clock he carried. He turned in the clock at the end of his shift. Again to prove the beach walk had been made.

Each man had several Coston flares which burned with a red glow and could be seen at night some distance. If a vessel that was coasting came too close to the beach, a surfman would burn a Coston signal to warn the vessel off. If a vessel in distress was spotted, a Coston signal was burned to notify those on the ship they had been seen and help would be coming. When a number of surfmen were gathered near the water when their was a rescue attempt, each of them would burn a Coston to illuminate the beach and show the place for rescuers to land with survivors. At first surfmen on beach patrol did not carry lanterns. This could be a hazard for them, since in fowl weather they often had to walk along the crest of the dunes back of the beach to avoid the high surf. The footing could be twisting and narrow, leading to a tumble down the face of the dune and into the surf. There were several spots like this on Cape Cod, Nantucket, Plum Island and other Massachusetts locations. The beach walk on a rocky shore near Annisquam, Straitsmouth, Nahant and Hull could be even more treacherous. And remember, you were by yourself in places where few people lived and on your own. In fact the very

The beach patrol was a major innovation in lifesaving that the service began in 1872. Surfmen walked the beaches from dusk until dawn every day and walked the beaches in daylight during stormy weather. Each man set out usually in opposite directions.

Coston flares were carried by surfmen on patrol and in groups to illuminate rescue scenes. They were ignited to show a troubled vessel it had been seen and help was coming.

isolation of some lifesaving stations was one reason the personnel turnover in those places was great. Monomoy Point and Knobbs Beach, south of Newburyport, Gay Head and the Nantucket stations come quickly to mind. In fowl weather the beach walks were continued in daylight hours in addition to manning the tower on lookout, watching passing ships was always a daily activity.

Surfmen had to master reviving drowned people. Their weekly drill on techniques was held on Friday. Each surfman had to take a turn at playing victim. Today it is called CPR.

The routine during the active season never varied from the practice the surfmen did. Once a week there was a surfboat drill, taking the craft on a carriage and launching it, rowing out and then beaching it. A more complicated maneuver was preparing the beach apparatus to fire a line to rig a breeches buoy for a rescue. This was done on two days. The boat house door would be opened and the men would each have a rope attached to the apparatus that would be hauled close to the water where a set-up would be made, using a sand anchor, a gun to fire the projectile carrying a line to a drill pole, then a hawser, a breeches buoy, and a wooden crotch which in actual practice would lift the line above the surf. This routine had to be completed in a set minimum time or the whole operation had to be repeated until the standard was met. On another day the surfmen practiced using a pair of flags, training with an international code. One day was for practicing first aid for the injured and for drowning victims. Surfmen would take turns being the injured victim or the drowned. Saturday was housekeeping with brooms, mops and pails and policing the area, picking up debris. It was a six-day routine. Sunday was for religious services and for the arrival of visitors, usually the families of the surfmen. Since one requisite of being chosen surfman was to be a resident near the station, their families usually did not have far to travel. The keeper's family often lived at the station. Sundays in warm weather were often visitors' day with the crew putting on demonstrations of rigging the breeches buoy, which young women from the neighborhood loved to ride. Launching the surfboat was frequently part of the show. It was often a picnic day, too. The training routine kept the men sharp and ready to respond to an emergency. That was the principal reason. It also filled their weekdays and Saturday with activity and helped avoid boredom, since it was often long periods of time between rescues.

The Number One surfman was the second in command at a station and took over the keeper's duties when the keeper was not there. If the keeper was absent for any length of time his substitute had his pay increased to that of the keeper. The Number One surfman was in many cases training to be a keeper, and when an opening occurred elsewhere, the District Superintendent frequently looked at his various Number One candidates for the promotion.

The keepers and surfmen had to have a thorough physical examination annually. For the surfmen it had to be at the end of the inactive season just before they were to return to duty. If a man failed the physical, he was dropped from the service. No matter how long or how well he had served, his lifesaving career was over. There was no pension system, no health care and no disability pay. If a surfman or keeper was injured or so weakened by his labors that he had to leave the service, he could under some circumstances receive his salary for up to one year. It depended on the man and his record; in most cases he got nothing. General Superintendent Kimball worked tirelessly to get the Congress to approve a pension for lifesavers but without success. Finally in 1915 when the Revenue Cutter Service and the U.S. Life-Saving Service were joined into the new Coast Guard, the lifesavers were given military rank and a pension. For the life of the Service they had monthly pay, ten times a year and all the dangerous risks that came their way. It was an unfair system.

Massachusetts played a major role in developing the ordinance used to rig the breeches buoy. In such use there had to be a way to throw the line to a wrecked vessel off shore. In Europe their system was to use rockets with a line attached. By 1845 the Massachusetts Humane Society had mortars to loft a line to a ship in distress. The British with their famous Royal National Lifeboat Institution was using the Manby mortar of brass with a five-and-one-half inch bore. Using a heavy charge of black powder the Manby tossed a projectile carrying a line a distance of three-hundred yards. That was too short for American rescue work. If the amount of powder was increased the shot line parted. There had to be a better way.

The next step was to use a cast iron Parrott Gun, a smaller version of Union Civil War artillery. Another attempt was that developed by Edward S. Hunt of Weymouth, Massachusetts, working closely with the Massachusetts Humane Society which adopted it to replace mortars at its stations. The Hunt Gun was purchased by the U.S. Life-Saving Service for some of its stations, but not all. There was a need for a line-throwing device of greater range and the Service found an ordinance officer to design it.

He was Captain David A. Lyle of the United States Army, a graduate of West Point and M.I.T. Lyle, for whom the gun would be named, began experimenting at the famous Springfield Armory in western Massachusetts. The search by Sumner Kimball for such a device began in 1877 and led to David Lyle. After some months at Springfield where the test neighborhood was too crowded, Lyle moved his experiments to the sands of New Jersey at Sandy Hook where there were few people and fewer structures. Within one year David Lyle had perfected his gun. It was made of bronze and weighed

The breeches buoy was used to rescue those on vessels wrecked close to shore.The gun to fire the line is on a carriage to the right. The rescued, including women and children, are huddled together in the center.

Surfmen often pulled their surfboats great distances down a beach to launch opposite a wreck. The keeper is leading the way with a Coston flare in this picture from a weekly newspaper in the 1870s when lifesaving got heavy media attention.

one-hundred and eighty-five pounds. Its carriage was wood with four handles, two on each side, for portability. The barrel of the cannon had a two and one-half inch bore and fired a bullet-shaped projectile a little longer than a foot with an eye bolt facing out where the line was attached. Lyle's gun threw the projectile and a line as much as two-thousand feet. Sometimes it was less than that, but enough. Sumner Increase Kimball was impressed and satisfied. The government adopted Lyle's cannon and manufactured them at the Springfield, Massachusetts Armory. Every one of its Life-Saving stations got one and some of the busier ones got two.

So the cannon to fire a projectile and line for the breeches buoy was developed and the procedure practiced twice weekly to make all lifesaving crews proficient at its use, but the important thing to remember is on the coast of Massachusetts by far most of the rescues and attempts at rescues were made by surfboat. The most important and most utilized piece of equipment at the thirty-two stations of the U.S. Life-Saving Service and those volunteer stations of the Massachusetts Humane Society was the surfboat designed for launching from beaches and later landing through usually heavy surf.

The Massachusetts surfboat was a descendant of small craft, oar-powered and used by Atlantic Coast fishermen. Those were its origins. They were made of wood and clinker-built and ranged in size from twenty-three feet in length to twenty-seven feet. For safety they had a good beam between seven and a half-feet to near ten feet for the longer ones. They were double-ended, meaning the bow and stern were spoon-shaped and were identical. They were light enough to be handled by a small crew, six men, and they weighed anywhere from seven-hundred and fifty pounds to half a ton. They varied in size and design to suit local conditions. All Massachusetts surfboats could be easily pulled through soft sand on a boat carriage to the edge of the water and launched quickly. Steering was by a man in the stern, using a sweep oar from eighteen to twenty feet long. Usually he was the keeper and from that position he could shout instructions to his crew, sometimes using a small megaphone. These surfboats were light draft and easily maneuvered in shallow water. After a while they were self-bailing, but not self-righting and their crews had to know exactly how to right them quickly in a strong sea as their lives often depended upon that.

Launching these surfboats in heavy weather often meant in a determined fashion just plain ramming them into a series of high waves breaking onto the shore. Ram straight ahead accompanied by hard pulling on the oars in unison! Surfboats would sometimes be thrown back to the beach repeatedly with the crew and keeper jumping out, waiting for a slight lull between combers, then trying again by pushing the surfboat, jumping in followed by hard rowing again. After many attempts, it sometimes worked. If there were too many failures, the keeper would wait for calmer seas. These surfboats were also equipped with a small sail that was stowed aboard and used to help the work of rowing the longer distances, especially against the tide. When sails were raised, a centerboard was lowered amidships to help steady the boat in a seaway. The centerboard could also be grabbed to help right the small craft if it overturned.

Lifeboats were much bigger and heavier, were self-bailing and self-righting and had to be launched from a carriage on a marine railway They weighed several tons and were usually from twenty-seven feet to thirty-five feet long.

Beginning in 1899 the Life-Saving Service started experimenting with twelve-horsepower gasoline engines in lifeboats. Despite the obvious advantages of engine power, adding that with its related equipment meant more than half a ton additional weight and a smaller carrying capacity of people. In 1906 the service was using twenty-five horsepower engines installed at the stern in the after air case with its fuel tanks at the bow.

Several of these were assigned to stations in Massachusetts. One was at Dolliver's Neck in Gloucester with another one at Monomoy Point. The Monomoy lifeboat was kept separate from that station at its own boathouse a mile north on the ocean with its own marine railway and a small crew assigned there all the time. Communication was by telephone. The surfmen were rotated from the others weeks at a time. The floating station at Dorchester Bay had one of the first power lifeboats on the East Coast and eventually added another. They did not have to be launched as they were in the water all the time.

Stations on the Great Lakes and the West Coast of the United States were the biggest users of motorized lifeboats. For the rest of the existence of the U. S. Life-Saving Service, that is, from 1906 until 1915 when it became part of the Coast Guard, Massachusetts remained surfboat country. Launch and row.

THE DAVID ATKINS TRAGEDY

There was loss of life during life-saving operations at Peaked Hill Bars, Station No. 7, on April 4, 1879 when the three-masted schooner, *Sarah J. Fort*, of Tuckerton, New Jersey on her way from Hoboken, New Jersey to Boston with coal stranded on one of the outer bars during a snowstorm in a northerly gale with a heavy sea. She carried a crew of six.

The vessel struck at one o'clock in the morning about a quarter of a mile off shore and was at that time unseen by the beach patrol. An hour later two new patrolmen walking in opposite directions came upon small wreckage and reported it to their station. Still later large wreckage and the outline of a ship were seen and reported to the station. Then Keeper David Atkins had the beach cart with a Lyle Gun start to the wreck scene with the other surfmen.

Atkins went ahead of the surfmen down to the shore opposite the wreck and thought the distance too great to be reached with the Lyle Gun, which was a new piece of ordnance just recently arrived. The surf was too high to launch a surfboat, so Atkins sent some of his men back to the station for the Parrott gun, a larger instrument which the keeper felt had a longer range.

From dawn until noon a total of nine shots were fired directly into the north wind without success, either falling short or the line breaking from the heavy powder charges. On the schooner the six men were still in the rigging when about mid-morning, with the ship breaking up, the cook fell from exhaustion and drowned and shortly afterwards so did the mate. By noon the tide had ebbed and the heavy seas had subsided at least

The Peaked Hill Bars Life-Saving Station was one of the first built on Cape Cod in 1872. It was a dangerous place for sailing ships caught in northeast storms. David Atkins was the first keeper here.

enough to attempt a launching of a surfboat manned by a picked crew from Peaked Hill Bars and others from the nearby station, No. 6, at Race Point, located a tad more than a half mile northeast of Race Point Lighthouse.

With Keeper Atkins at the stern on the steering oar the crew sprang to their places and started pulling through the surf, but almost at once a wave struck and filled her half full. The crew immediately dragged her back to the beach and rolled her on her side to empty her.

Atkins decided to wait about twenty minutes for a further ebbing of the tide in the hope the violent surf would subside. It did not, but a second launch was attempted with the surfboat clearing the first breaker but not the second which combed over her, filling her with water. At that point she was wallowing deep in the water and full to the gunwale when a huge wave picked her up and tossed her on the beach where five of her timbers broke open. The crew jumped out and hauled her from the surf, finding her damaged badly and unfit to use.

Keeper Atkins had asked Captain Harvey Cook of Provincetown who had arrived on the scene with others after hearing of a wreck to go back to town for a whaleboat. That craft finally arrived and was to be used as a substitute in yet another attempt to reach the four weary survivors on the schooner.

Surfmen in cork life-jackets launching a surfboat for yet another rescue attempt at Peaked Hill Bars.

One of the men who came to the scene from Provincetown because of the news of a wreck was Captain Isaac F. Mayo, a surfman of long experience. Mayo assumed command and quickly chose a fresh crew. He did not include Atkins or any men from his station who were tired and hungry. They had not eaten since the night before and had been working on the beach since three o'clock that morning in the cold wind and had twice been dumped in the bitterly cold ocean at the northern tip of Cape Cod.

Atkins and some of his men begged to be included, but Isaac Mayo said no and remained adamant. Mayo and his fresh crew of volunteers had to make two attempts at a launch with the whale-boat filling with water and then being emptied before the third try was successful.

The greatest threat to the rescuers was all that debris in the water from the schooner, *Sarah J. Fort* that had been breaking up for many hours in the heavy surf. There were timbers, deck planks, wooden railings, cordage, sails and other pieces of broken wreckage tumbling in the water and much of it concentrated near the remains of the vessel.

Captain Mayo cautiously moved his small craft near the schooner and one by one took the four men off. It was a particularly difficult rescue, since the sailors were exhausted and badly frost-bitten and the ship's captain was especially far gone.

Mayo carefully headed for the beach and safety in a following sea when on the last breaker a heavy wave swept under the stern, lifted the craft and tossed it end over end, throwing all of the men into the water with its deadly undertow. They were instantly in danger of drowning or being swept away in the surf when some of the spectators rushed into the water and helped them to safety.

The four men from the schooner were immediately taken to the Peaked Hill Bars Lifesaving Station where they were treated. Their frost-bitten hands and other extremities were dressed to help restore circulation, then they were wrapped in woolen blankets and fed hot food. Dr. Crocker, the port physician from Provincetown, arrived after he received a message from Keeper Atkins. Crocker helped restore the stricken sailors to good health. Two of the rescued men stayed at the station for four days. The other two remained a shorter period.

The greatest difficulty of this wreck and rescue was for Keeper David H. Atkins of the Peaked Hill Bars Station when the official annual report of the service was published it said:"It is to be regretted that Keeper Atkins's success was not equal to his efforts on this occasion. He has always been regarded as an efficient keeper, and his record for fidelity, courage, and energy at scenes of shipwreck, is known to the officers of the service. The Lyle gun would probably have established the desired line communication with the wreck, and his failure to employ it, through a strange inapprehension of its merits, must be deplored."

In the United States Life-Saving Service that was a devastating comment on a man's reputation. However, Atkins, a capable and proud man, was not removed from his post as keeper at Peaked Hill Bars Station.

Later Captain Isaac F. Mayo was awarded a gold medal for his work in directing the rescue of the four from the schooner *Sarah J. Fort* in that wild surf.

The Peaked Hill Bars Station was one of the original nine built in 1871 and placed in service the following year. Its name comes from the two lines of bars that are submerged and run parallel to the coast offshore for six miles. The inner bars are about four-to six-hundred yards from the beach and the cuter bars about fourteen-hundred. Not many ships strand on the outer bars, but they do sweep over them and come to grief on the inner bars. That area was an ocean graveyard for vessels and their crews.

This four-masted schooner is stranded off shore and in bad condition with wind damage to her sails and rigging. Her crew has already been rescued and she is abandoned to eventually break up.

The tides that surge around that tip of Cape Cod with great velocity range from eight feet to as much as twelve. In foul weather, Peaked Hill Bars has always been a treacherous location.

On the 30th of November, 1880, the sloop, *C.E.Trumbull,* of Rockport, Massachusetts, struck at four a.m. at that dangerous place in a rough sea with a crew of six. The *Trumbull* was a granite-carrying vessel whose Cape Ann homeport was near numerous quarries back of the village of Pigeon Cove. She was on her way home from New Bedford without cargo and with some coal as ballast. The vessel was seen by the beach patrol who took the word to the station where Keeper Atkins ordered his crew to run out the surfboat. Atkins told surfman John L. Cole to stay on the beach and keep two lanterns burning in that high wind. The rest of them along with Atkins made for the stranded vessel off shore, which when they arrived, was headed eastward the same direction as the strong but low tide. The wind was high, a gale, and the sloop was threatened by the conditions.

Atkins skillfully moved the surfboat under her stern near the lee side. He threw a line and it was made fast. With the high waves and the boom swinging, Atkins saw that they had to move rapidly and he shouted to the crew of the *Trumbull* to jump into the surfboat. Four of them did and the other two held back, evidently going for their clothes. Atkins yelled, "We didn't come here for your luggage! Get in the boat!" They did not. "Cast off!" shouted Atkins, but they did not. "Cut!" Atkins yelled, and they cut the line and the surfboat moved quickly away from the sloop's side. The surfmen pulled for the beach where they let the four men off and returned to the stricken vessel. The *Trumbull* had an extra-long boom for loading and unloading granite blocks and it was swinging wildly as the sloop rolled in the waves. The boom came up under the cork belt of the surfboat and flipped it over bottom up, throwing Atkins and his crew into the frigid water where they climbed on its bottom and rolled it over right side up. A huge wave rolled it over again and the surfmen climbed on the bottom and were washed off two or three times. Atkins shouted above the tempest, "Stay with the boat!" Some of them did, along with Atkins; others started for the beach, including Isaiah Young who had a tough swim for the shore as he headed for the burning lanterns.

Young lay on the beach unconscious until he heard John Cole, standing over him, say, "For God's sake, Isaiah, is this you?" Cole then picked up Young and carried him to the Peaked Hill Bars Station where he recovered. Young thought he stayed by the overturned surfboat a half hour before swimming for the beach. He had high praise for his cork life belt, saying it enabled him to reach shore. Cole went back to tend the lanterns.

Surfman Charles B. Kelley, one of the three survivors, guessed that he stayed by the surfboat twenty minutes, maybe thirty, after Keeper Atkins had shouted to his crew, "Stay by the boat!" Finally, in desperation Kelley struck out for the shore, stopping to kick off his hip boots and occasionally turning on his back to rest in the heavy surf while floating with the aid of his cork life belt.

When Kelley saw the powerful beam from Highland Lighthouse and the lanterns that Cole had on the beach, he swam toward them and finally landed two miles east of his station.

The third survivor, Surfman Samuel O. Fisher, told of seeing Keeper Atkins and one of his crew, Stephen F. Mayo, in the water near the overturned surfboat. A wave carried Fisher from the boat and he passed Taylor who was calling for help but making no effort to help himself. Mayo and Taylor did not survive.

When Fisher reached land he threw off his cork life belt and his heavy, wet coat, as they weighed him down, then he started to walk toward Cole's lanterns down the beach. Back at his station he received dry clothes and a shot or two of brandy.

After that he returned to the beach to search for survivors and after walking east toward the Highlands Station, he came across the body of his captain, Keeper David H. Atkins, who had been brought out from the surf by someone and carried above the high water mark.

Atkins had been discovered earlier by Surfman L. S. Small from Station No. 8, Highlands, who was walking his beach patrol westward after 4 a.m. on November 30, 1880. On his way back from the end of his beat, Small saw something black roll up in the white foam of the surf. Running down to the water, he was horrified to discover the form to be not only a man, but a lifesaver wearing a cork belt, a man who was Keeper David Atkins. Small tried to restore life to the body, and he worked on him until he was satisfied he was dead. It was Small who carried Atkins back above the high water mark and placed him on the sand, removing his cork belt and placing it over the face to protect it from the elements. On the way back to his station, Small saw another body out in the surf, but the tide and wind kept it offshore.

When he reached his station he reported his findings to the Keeper, E. P. Worthen, who then along with Surfman Lucas walked down the beach where they found Elisha Taylor in the surf and brought his body ashore.

Shortly after that the body of Stephen Mayo appeared in the water but too far out to be recovered. Worthen left Lucas to watch the body drifting while he returned to his station for a rope and grapnel.

After sending two more surfmen to help Lucas, Worthen asked the son of Keeper Atkins, who was a surfman at Station No. 8, to accompany him to the telegraph office to notify Benjamin Sparrow, of East Orleans, the District Superintendent of the Second District, Coast of Massachusetts. Walking with young Atkins, Keeper Worthen told his companion that it was his father whom Surfman Small found in the water and that probably the entire crew from Peaked Hill Bars was gone. Worthen said, "Go home and tell your mother, son. She will need you now." Young Atkins walked away toward his home, wondering how he would be able to tell her.

After wiring Superintendent Sparrow of the tragedy, Worthen walked out to the bank and saw that the body of Stephen Mayo hadn't been recovered. Then Worthen went to the beach with the help of surfmen Lucas and McKinnon of his station, Highlands, and the three of them followed it to a spot one mile south of Highland Light where they finally brought Mayo's body through the surf and onto the shore. A team of horses was used to bring the body back to the station at Peaked Hill Bars.

If the captain of the sloop and the pilot, the other two of the six-man crew, had jumped into the surfboat instead of going for their luggage, all of them would have been rescued in one trip; Keeper Atkins and surfmen Taylor and Mayo would not have died.

After the surfboat capsized, throwing the lifesavers into the water, the sloop slued off the bar and drifted south where she was seen the next morning bouncing in the surf off Chatham where the captain and pilot had hoisted a flag of distress. A wrecking party came aboard and sailed her to an anchorage. The sloop *Trumbull's* damage was mostly to her sails, not her hull. After acquiring a new jib and three men for a crew, she sailed off to Boston, leaving behind three dead lifesavers.

Three days later, December 3rd, 1879, was a bright winter day and Provincetown was in mourning. Its citizens were drawn to the largest church in the village and they were wearing their Sunday best clothes. There were at least fifteen hundred of them crowding the spacious interior of the Centre Methodist Episcopal Church and many outside unable to get in.

Business was suspended, and at the wharves, and on board the vessels in the harbor, the colors drooped at half staff. From Monomoy in the south to Race Point the United States flag and the district flags were also at half staff.

Superintendent Benjamin Sparrow was there with members of his lifesaving crews from Cape Cod Stations No. 6, 7, 8, 9, 10 and 12. Those men were the pallbearers who slowly carried the three coffins into the church through the huge crowd who had come to say goodbye to the lifesavers.

The ministers of different denominations had been invited to take part in the service and they spoke to comfort the families of the men and praise their gallantry in their dangerous duty.

At the end of the service the three crippled survivors-Isaiah Young, Charles Kelley and Samuel Fisher-bent over the coffins, shaken with emotion, taking in silence the last leave of their fellow comrades who had shared with them so many desperate adventures off that coast.

The surfmen, acting as pallbearers, slowly carried the coffins out to the hearse and the crowd followed the horse-drawn vehicle to the cemetery. A brief prayer and benediction followed and the coffins were lowered into the ground.

A place like Provincetown, a village by the sea, had a special feeling for the men of the U.S. Life-Saving Service and the duty they performed. It was no different here than in other coastal places where men had gone down to the sea in ships since time began. Every neighborhood along these shores had lost men at sea. The people who stayed home were conscious of the dangers to those who went out on the water and the brave role the lifesavers played in going out to rescue those in peril no matter the danger to themselves. Their exploits were told and retold in the nearby towns and villages. It was part of history and lore. An important part.

When he died, Captain Atkins was forty-three years old, a water man who had spent much time in nautical pursuits-fishing, whaling and coasting. He had been for more than ten years a volunteer custodian of a local Massachusetts Humane Society boat house. When the Life-Saving Service built the Peaked Hill Bars Station and eight others on Cape Cod in 1872, Atkins became the first keeper at the Bars, a position he held until his death that November night in 1880.

For twenty months Captain Atkins had lived with what was considered his failure to use a new Lyle gun at the wreck of the schooner *Sarah J. Fort* and then had to endure the mortification of seeing the rescue by a volunteer crew in a fresh boat brought from Provincetown.

The proud and sensitive man felt the shower of comments of rough stings no less keenly for the silence with which he bore them. His superiors knew him to be a gallant leader, a fearless man, and they signaled their opinion by retaining him as keeper at Peaked Hill Bars Station.

At the beginning of the active season that year, when he left his home in Provincetown, his wife cautioned him to be careful. Atkins replied, "I will endeavor to be so, but you must fully understand I will die at my post before I would either disgrace myself or the Service to which I belong." It was in that spirit that David H. Atkins faced his duty, and met his death.

He was not well off; and he left his widow with two small children. Surfman Elisha Taylor was a poor man, who was survived by a widow and several tiny children. Stephen Mayo was single and the only child of elderly parents with a father in poor health who relied upon him.

Of the three survivors, both Surfmen Kelley and Fisher who were severely injured had a long convalescence before recovery. The other surfman, Isaiah Young, nearly died of his injuries: cutting pains in his chest from hemorrhaging in his lungs and one leg paralyzed, a cripple for life.

There were no pensions in the Life-Saving Service. Injured members were simply cut loose, unless they could pass the strict physical exams to remain on the payroll.

However, there were public collections held in Massachusetts with the widow of Atkins receiving eleven-hundred dollars. The family of Elisha Taylor got sixteen-hundred dollars, and the aged father of Surfman Stephen Mayo, six-hundred dollars. Eight-hundred dollars went to the badly disabled Isaiah Young.

At the Peaked Hill Bars Station there were all new faces, except one, John L. Cole. Superintendent Sparrow had to move quickly to put new staff in that difficult location.

The first night the new crew was there, the beach patrol started walking, from dusk until dawn, one surfman each way, to the east and to the west, looking for ships in trouble along the most dangerous winter coast in the world.

HORROR AT DEVIL'S BRIDGE

One of the most unnecessary sea tragedies in New England took place on a bitterly cold night in January, 1884. The passenger steamer *City of Columbus* of the Boston and Savannah Steamship Company on one of her routine voyages between Boston and Savannah strayed off her course and struck boulders off the northwest tip of Martha's Vineyard at Gay Head. The accident happened at about 3:45 in the morning. The captain who had been lying down in his cabin adjacent to the pilot house ordered the engine room to back hard and the ship did so, filling with water and sinking. Of the 132 people on board only 29 survived.

On all the charts of Vineyard and Nantucket Sounds the spot where the City of Columbus came to grief is named Devil's Bridge, an appropriate name in view of what happened there.

It was a night of fear, horror and frozen bodies. The ship leaned heavily to port and many passengers drowned in their staterooms. Some of those who reached the decks climbed the ratlines to the rigging from where many of them survived and others died from the cold and from drowning after falling into the frigid waters. Still others who managed to reach the deck were swept away by the huge waves. The worst statistic of that early morning of January 18 is this: one-third of all those aboard were women and children, none of whom survived the wreck of the *City of Columbus* on Devil's Bridge.

The deck or navigating officers of the *City of Columbus* were all Cape Codders. The Captain, Schuyler E. Wright, was born in Wareham Village on Buzzards Bay. He was 52 years old and had spent 39 years on the water, having gone to sea at 13. He was a master and licensed pilot. The *Columbus* was his fifth command. Wright exemplified a devotion to duty. He never removed his clothes during the three-day trip between Boston and Savannah and the way north. His stateroom was right next to the pilothouse and the door between the two was never closed.

The First Officer was Edwin Fuller of Barnstable. He was also 52 and a licensed master and pilot. He, like Captain Wright, had gone to sea as a boy.

The Second Officer was 27-year old Augustus Harding of Chatham who had never gotten his pilot's license, although he had spent some years on other steamers before he signed on three years before to the *City of Columbus*. Captain Wright respected Harding, calling him "a good pilot." Harding's father was a Cape Cod shipmaster. Captain Wright lived through the tragedy; Fuller and Harding did not.

Of the engine room officers, two survived and two did not. The survivors were First Assistant Engineer, H. A. Phillips of Boston, and Second Engineer, Henry Collins of Taunton who was brought ashore in the first Gay Head lifeboat. The Chief Engineer, Archibald Morrison and William Murray, third engineer, both from Boston, died with the ship.

Many in the crew had spent years going to sea and they had a vast amount of experience. Almost all of them were either from Massachusetts, Maine or the Canadian Maritimes.

The *City of Columbus* was five years old when she died on the rocks of Devil's Bridge beneath the red and white beams from Gay Head Light above her on Martha's Vineyard. She was built on the Delaware River at Chester, Pennsylvania by the John Roach & Son shipyard. She had three iron decks on an iron hull with wooden superstructure and deck houses. The one stack was amidships between heavy fore and aft masts that carried sails to steady her in a seaway. She was 275 feet in length with a 38 foot beam. The 1500 horsepower compound reciprocating engine had a four and one-half foot stroke that turned her propeller 62 revolutions per minute, giving the vessel a speed of 12 and one-half knots. The *Columbus* weighed 2,200 tons and had a passenger capacity of 200 in her 42 staterooms and in the separate bunk rooms for men and women in steerage forward under the bow on the main deck. Safety equipment included six metal lifeboats and a life raft. Every berth had a life preserver. In addition to passengers the *City of Columbus* and her sistership, the *Gate City*, each had a cargo capacity of 2500 tons which often consisted of shoes, furniture, fish, bacon, pianos and assorted products manufactured in New England. Coming north the cargoes were cotton, rosin, potatoes, oranges, and watermelons. These vessels were profitable and her owners, the Boston & Savannah Steamship Co. had insured the $300,000 *City of Columbus* for $250,000.

The *City of Columbus* at Nickerson's Wharf on Fort Point Channel in Boston.The fatal trip began here at 3:13 on the afternoon of January 17, 1884.

It had been more than four hours since the survivors from the *City of Columbus* had climbed into her rigging to escape the sea crashing over the stricken vessel. Suddenly from the west there appeared a chance of rescue. A steamer was approaching and surely

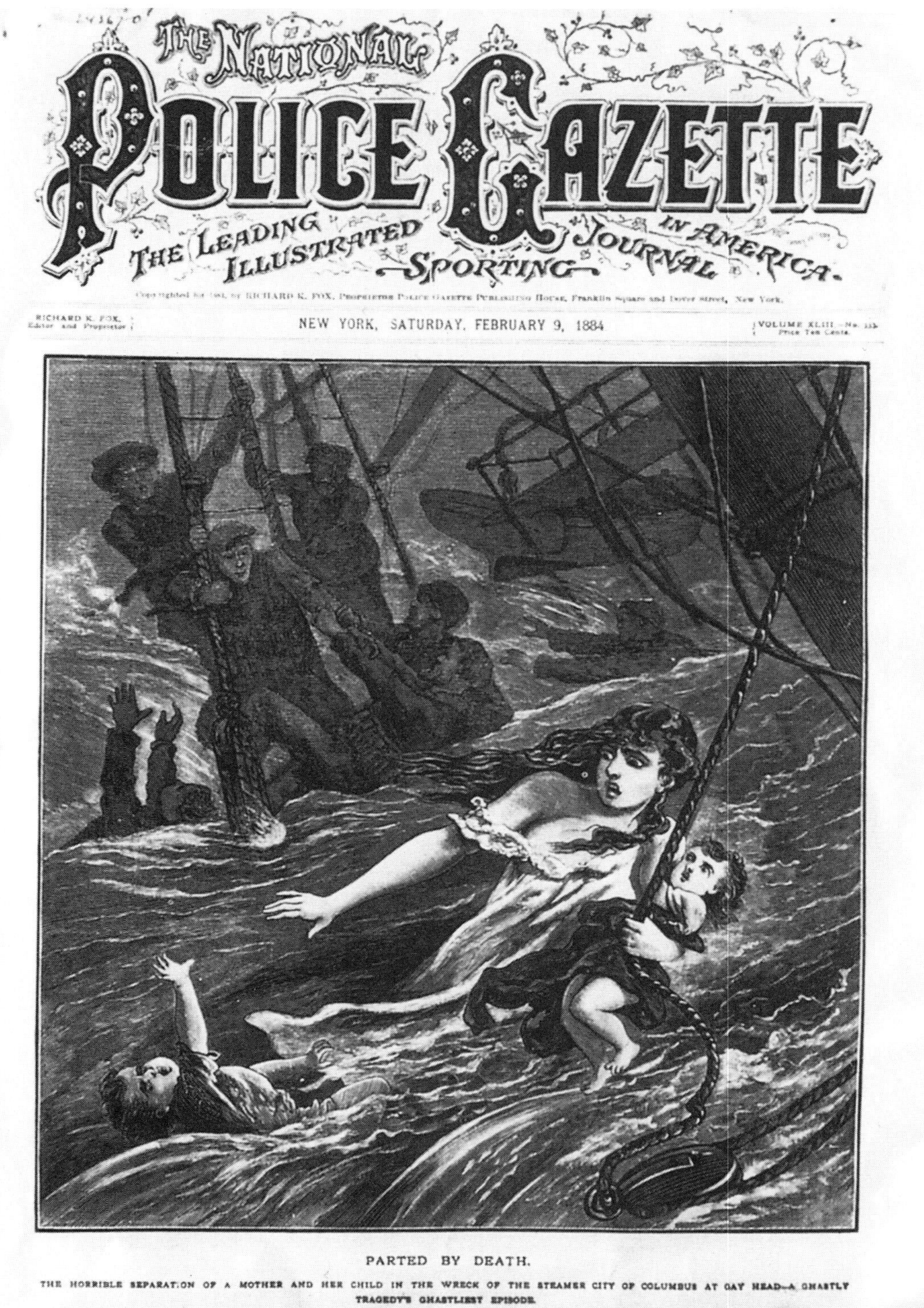

THE NATIONAL POLICE GAZETTE

THE LEADING ILLUSTRATED SPORTING JOURNAL IN AMERICA.

Copyrighted for 1884, by RICHARD K. FOX, Proprietor Police Gazette Publishing House, Franklin Square and Dover Street, New York.

RICHARD K. FOX, Editor and Proprietor | NEW YORK, SATURDAY, FEBRUARY 9, 1884 | VOLUME XLIII.—No. [illegible] Price Ten Cents.

PARTED BY DEATH.

THE HORRIBLE SEPARATION OF A MOTHER AND HER CHILD IN THE WRECK OF THE STEAMER CITY OF COLUMBUS AT GAY HEAD—A GHASTLY TRAGEDY'S GHASTLIEST EPISODE.

An artist's sketch of a desperate mother watching in horror as one of her children is swept away from her on the deck of the *City of Columbus.* Women and children were one-third of those aboard. None of them survived.

it would see them. Sunrise was at 7:23 and it was past that time, at least half an hour past. It was morning bright and the wind had turned to the northwest, making for even colder temperatures. There was hope alive in the masts. The steamer approaching, heading east was the *Glaucus* out of New York for Boston, her regular run. Across Vineyard Sound in the pilothouse of the *Glaucus* was Captain Maynard Bearse, a veteran with 14 years as a pilot and 12 years a master in steam. Bearse had spent nine years as captain of the *Glaucus* on this run. He knew the schedules of the regulars that sailed around the Cape and over Nantucket shoals and Vineyard Sound. The *Glaucus* was sailing close to the Elizabeth Islands more than four miles across the water.

Henry Collins, the Second Engineer of the *Columbus,* from his perch in the rigging started waving his cap before the other steamer was abeam of the wreck on Devil's Bridge. Later Collins said he thought that he waved the cap wildly, swinging his arm for at least 20 minutes. Another survivor in the rigging, Captain Thomas R. Hammond of Gouldsboro, Maine, was higher up in the maintop. Hammond said that he clearly saw men standing along side of the pilot house of the *Glaucus*. He thought that the men in the rigging around him in dark clothes and white life preservers should have been equally visible to those on the *Glaucus*. However, the steamer did not deviate from her eastward course.

Still those in the pilot house of the other steamer had been watching the wreck closely for some time, maybe as much as half an hour. Bearse was looking across the Sound through his glass and knew it to be the *Columbus* from the position and shape of her stack. Bearse knew the *Columbus* was due past the Vineyard by about four a.m. and that was at least four hours before, he later testified at the hearing about the disaster. He told the inspectors at the official inquiry that he could see the masts and the waves breaking over the vessel. There was a heavy fog bank along the shore that obscured the rigging, he said, so he never saw the men in the rigging, which was strange considering what Captain Hammond said. "I talked with the officers, and we supposed the people must be all lost or all ashore by that time," said Bearse at the inquiry. There were two passengers among the six people in the pilot house and Bearse said "no one suggested that we deviate from our course to see if we could render help. It would not have detained us long to see if anybody was on the vessel."

His testimony is astonishing because he said "no one suggested that we deviate" as if the others present were in charge and Bearse, although he was the Captain, was just another observer. No, Bearse was in charge, so the decision to investigate the wreck was his. The other comment about being detained does refer, I believe, to the thinking by ship captains about the pressure from owners about maintaining schedules. The word from ship's owners to captains was "don't be late, it's expensive." Bearse's failure to sail across Vineyard Sound and investigate the wreck on Devil's Bridge could have been the decision that wrecked his career and the careers of his mate and his chief engineer who were with him in the pilot house that fateful morning.

The newspapers editorialized against the captain and the other two officers who were topside on the *Glaucus* and continued to sail eastward without investigating the wreck.

Volunteers of the Massachusetts Humane Society rowed out to the wreck and rescued survivors. The men who manned the society's lifeboats were Gay Head Indians from that area of the island. The wreck was spotted after dawn that morning by

Alice Atkinson of Woodstock, New Brunswick, was a member of a family tragedy. The 30-year old woman died as did her son, her parents and her nephew.

Below left: Alice Atkinson's son, Wilmont L., was eight years old when he died on the *City of Columbus*. This family photograph of him as "The Little Corporal" with his sword and his ice skates was taken when he was six.

Below right: Richard C. Merrill, a five-year old from Skowhegan, Maine, was a nephew of Alice Atkinson. His body was found floating in Vineyard Haven Harbor. The medical examiner described him as "one of the prettiest children I have ever seen."

Wampanoag Indian surfmen from Gay Head are those in the Humane Society surfboat they rowed out two and one-half miles to rescue eight people from the wreck on Devil's Bridge. At the steering oar is Joseph Peters, a veteran whaler, waterman and longtime volunteer lifesaver.

Below: With the return of the first Gay Head lifesaving crew and their obvious exhaustion, a second crew of fresh men was assembled with Joseph Mosher of Chilmark in charge at the stern. These men rescued thirteen people at the *City of Columbus.*

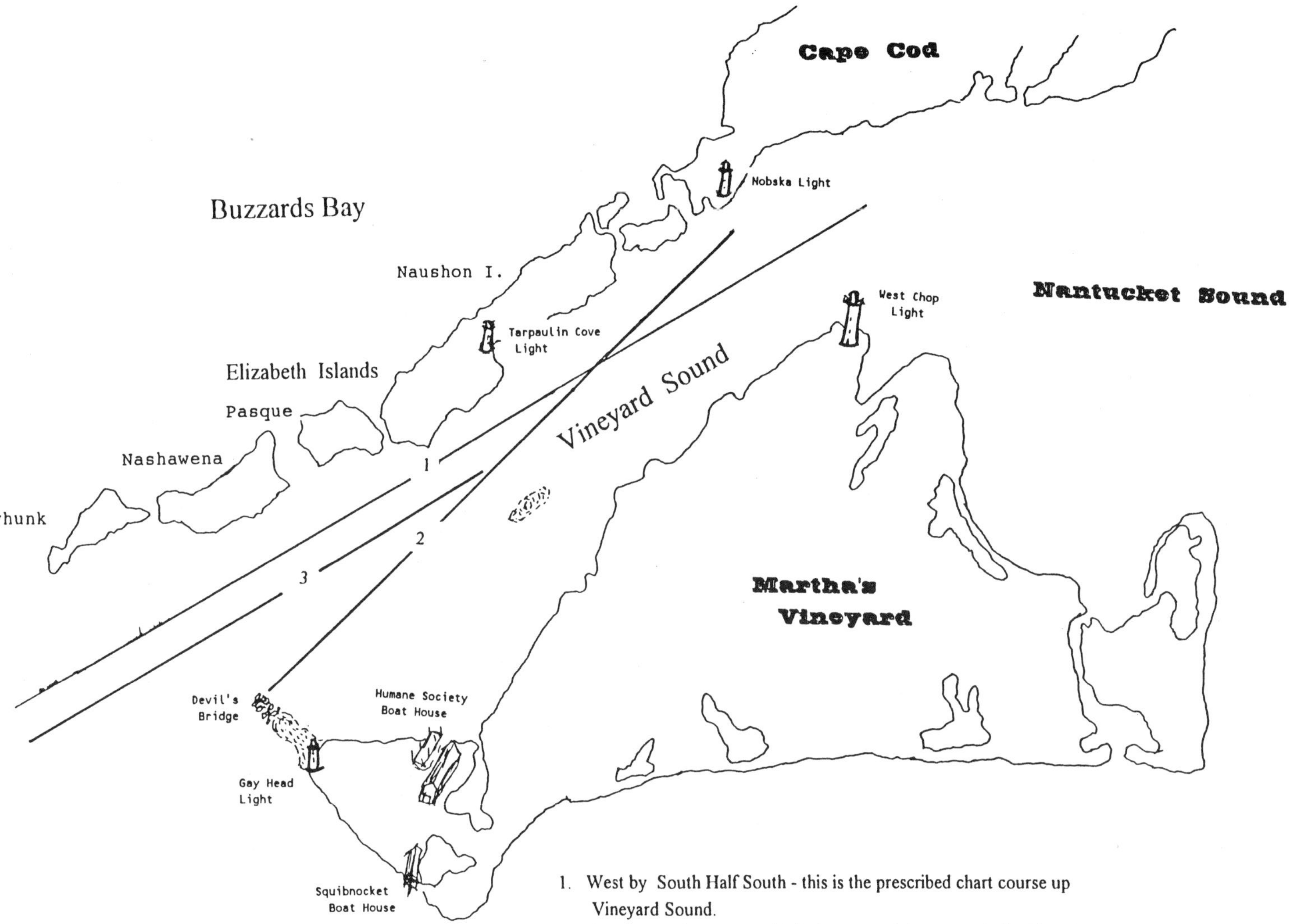

1. West by South Half South - this is the prescribed chart course up Vineyard Sound.
2. Southwest by West - this is the course it is accepted the City of Columbus steered to disaster.
3. West Southwest - this is the course change which was ordered but not made good when Tarpaulin Cove Light bore north.

Horatio N.T. Pease, the keeper of the Gay Head Lighthouse on the bluff above the scene. Pease was also in charge of the Humane Society's local lifeboat. When he spotted the masts and the slightly exposed bow of the wreck, he sent word to nearby houses and to the village of Squibnocket, some five miles away.

There was a strong west wind and a heavy sea running and it took more than an hour for six men to pull the lifeboat westward to the wreck. Their rowing against the wind and sea exhausted them so that when they reached the mostly sunken hull they could do nothing but hold on for awhile to halyards hanging from the forward mast into the water. The waves had torn away the deckhouses and there was so much loose wreckage along side of the *City Of Columbus's* hull that the lifesaving crew could not get close enough for a direct rescue. The Indians in the lifeboat shouted to those in the rigging to jump into the water where they would be picked up.

Seven men in the rigging jumped clear of the hull beneath them and into the water. The first was a fireman who vanished from view and after reappearing was reached with a boathook and pulled to the lifeboat and lifted out by eager hands to safety. The other six followed, jumping into the water, and then the volunteers signaled that they had a boatload of six crewmen and one passenger and would return. The long trip back to land further exhausted the Indians, one of whom had responded so quickly to the emergency that he had forgotten to put on shoes.

Back on land keeper Pease had gathered another volunteer crew of six Gay Head Indians. They took the lifeboat from the first crew and started out for the wreck where the rigging held so many humans that where they were clustered the space seemed to be black with them. By this time, after 10 o'clock in the morning, more frozen, exhausted men had dropped to their deaths. With the second rescue trip on the way, the stack of the *City of Columbus* with waves breaking around and over it began to shake and it was clear it, too, would soon break loose as the deckhouses had. A passenger, Captain Thomas R. Hammond, of Gouldsboro, Maine, was in the rigging and saw the danger to himself and others huddled aloft from the swaying smokestack, which he felt would soon tear loose. Hammond, a master mariner, moved to a point where he had the forward mast between himself and that shaking stack. He encouraged others to do the same. Some of them did, but when the stack broke loose pulling the stays with it, a passenger in the shrouds was carried to his death. It is not known how many of those who reached the rigging fell to their deaths before help arrived. There was more help on the way.

After he had rounded up his first volunteer crew, Pease sent to Squibnocket for the Humane Society's other lifeboat at the shelter there. It was moved over land by another volunteer crew of six Gay Head Indians who, using horses, took it to a launching site as close to the wreck as they could. Once there, they entered the water and started pulling toward Devil's Bridge.

A young master mariner was a passenger heading south to take command of a sailing vessel for a commercial voyage. He was Sherrington Vance, a 25-year old captain from North Truro, Nova Scotia, whose destination was Pensacola, Florida where he was to set sail for South America. Vance worked himself up to the main deck after the *City of Columbus* crashed onto the boulders of Devil's Bridge and sank. Waves were breaking over the bow and cascading over much of the hull and dissipating toward the rear mast, which was the only part of the vessel above the water toward the stern. Vance was struck by a comber that swept him off the deck and near a lifeboat that was loaded with water and drifting not far from the shipwrecked vessel. Vance swam toward the boat and grabbing the gunwale pulled himself in. At that moment the lifeboat turned over. Vance knew he had to right it. "It was a fearful struggle, but it was for my life and I knew it," he said.

A navy coastal tug, the *Speedwell*, was anchored in Vineyard Haven Harbor where she had stopped on a voyage from Washington, D.C. and Norfolk to Portsmouth, New Hampshire, carrying freight. About 11:30 that morning the Collector of Customs came alongside and asked the tug to head for Devil's Bridge for rescue work. On the way west several bodies were seen floating and some wreckage of a deckhouse. Farther on in front of the tug appeared a lifeboat filled with water and barely afloat. The *Speedwell* stopped and launched its own small boat which was rowed over to the drifting one. To their surprise they found an unconscious man. It was Sherrington Vance. He had been in that lifeboat almost ten hours. Inside the warm tug Vance was revived. The *Speedwell* continued toward Devil's Bridge where a Revenue Cutter, the *Samuel Dexter,* was anchored. Around the wreck were the two Humane Society lifeboats and small boats from the *Dexter*.

There was nothing for the tug to do at Devil's Bridge that afternoon so she headed back for Vineyard Haven, only stopping to pick up in the water three bodies from the *City of Columbus* along with an overcoat, one box and four trunks. The *Speedwell* made her way to a dock where several crew members helped Sherrington Vance walk up the wharf. He felt weak, but the arm of a sailor on each side enabled him to know that he would make it into town.

The first lifeboat from the *City of Columbus* carried seven people. It was moving away from the stricken steamer when the 35-year old quartermaster, Roderick McDonald, leaped toward it from the deck and hit the water instead. McDonald went under then surfaced, grabbing the gunwale. He shouted for help, but those in the lifeboat ignored him. A massive wave rolled in, nearly filling the lifeboat. A stewardess-Miss Teresa J. Smith, of Boston-and two men panicked and jumped into the water, trying to swim back to the wreck for safety. All three drowned; her body was never found.

In those long, rolling swells McDonald waited for a wave to lift him. In that brutally cold water his patience finally paid off. The right wave came in and with its help he pulled himself over the side of the metal lifeboat and tumbled in. Above the scene some 170 feet from the water the Gay Head Light with its first order Fresnel Lens continuously flashed three white and one red with ten seconds between each.

Down the cliffs from the lighthouse the breakers were pounding the rocks and McDonald wanted to avoid a landing there. Far too dangerous, he thought. Taking command he used an oar to keep the metal lifeboat moving east parallel to the shore, always looking for a good place to land.

The small boat was heavy with water. There was some bailing, but it did little good. The boat had been holed when launched. The others rowed and as they approached the shore two miles east of the *Columbus*, one man, a passenger, died at his oar. When the lifeboat touched the beach the men, too weak to walk, fell out and then crawled on

Another group of rescuers were these Gay Head Indians who launched this whaleboat into the strong surf where it was holed on rocks in the shallows. The crew jumped out and pushed the boat back to the bottom of the cliffs where the next day this picture was taken.

hands and feet. McDonald reached a house and sat by a fire with the Indian family whose place it was.

James Brown, who had been a steerage passenger, fell unconscious on the beach not far from the boat. He had gone blind Hours later-he didn't know how many hours-he woke up to see a Wampanoag Indian woman leaning over him with a container of hot rum and ginger. He knew he was in a house, but he had no idea how he got there. It was warm. When he saw the steam rising from the drink she offered him, he was aware that he could see again. More than that James Brown also knew that he would live.

The first volunteer crew for the Humane Society's surfboat was in charge of Joseph Peters, the others were Moses P. Cooper, James F. Cooper, Samuel J. Haskins, Samuel J. Anthony and John P. Vanderhoop. All of them were from Gay Head.

With the first boat underway, five other volunteers in an effort to help in the rescue tried to launch a whaleboat from the beach. They tried to avoid the rocks and get to open water off shore, but the heavy surf grounded them on the boulders, tearing holes in the boat, filling it with water. The crew managed to reach shore safely, drenched and bitterly cold. The whaleboat was a total loss. The crew was Thomas Jeffers, Henry Jeffers, John Lula, Simeon Divine and Charles Stevens.

When the Society's boat returned from its first trip, a second volunteer crew was waiting. From Chilmark there was James T. Mosher; the others-Leonard L. Vanderhoop, Charles Grimes, Patrick L. Devine, Thomas C. Jeffers and Peter Johnson-were from Gay Head. The second trip with a new crew was out on the water for three and one-half hours. During that time they rescued 13 people and took them to the Revenue Cutter *Samuel Dexter*. The first time at the *City of Columbus* they picked up five people. They returned to rescue six whom they took to the *Dexter*. Their third trip saved the last two. The two Humane Society volunteer crews were officially credited with saving the lives of 20 of the 29 survivors.

The *Samuel Dexter* was a revenue cutter of the Revenue Marine Service, a federal agency responsible for enforcing customs laws and for protection of the coast. She had been on patrol and was passing Devil's Bridge around noon when someone on board spotted the masts of the *City of Columbus*. The *Dexter* moved closer where the identity of the stricken vessel could be determined. Only the bow rail was visible forward and that often was obscured by huge waves pounding her starboard. To those on the *Dexter* it was a dangerous situation with the wreck inside the Devil's Bridge buoy. The Captain of the *Dexter*, Eric Gabrielson, seeing survivors clinging to the rigging ordered her anchor dropped. The *Dexter* was off the starboard side of the wreck to the north. The second Gay Head surfboat was tossed constantly in those conditions described later by Gabrielson as "the wind blowing very hard, the sea rolled mountain high." A seaman was ordered to stand by the line with a slip rope to let go the anchor if necessary.

The United States Revenue Cutter *Samuel Dexter* was the rescue steamer of the disaster. She did coastal patrol for law enforcement with customs and seacoast protection. The *Dexter* was built in Boston in 1874 and was one of the last in that service to have a wooden hull. She is shown here in New Bedford Harbor, her homeport.

The Captain ordered the *Dexter's* cutter be made ready to launch and Second Lieutenant John U. Rhodes to be in charge. Five members of the crew volunteered to go with Rhodes. In that turbulent water the *Dexter* rolled, rails under, with her crew holding on to anything solid and unmovable to avoid a fall or being washed overboard.

Once they were underway through the huge combers the lieutenant and his rowers worked their way toward the *Columbus*, but were wary of getting too close because of the wreckage around the bottom of the masts. Rhodes shouted to the men in the masts to jump during the lull between waves. It was mid-afternoon and the survivors had been in the rigging since about four o'clock that morning, close to 12 hours. Some of them were frozen to the rigging. But they began to jump into the water where they were reached quickly and pulled into the small boats. The crew from the *Dexter* picked up seven men this way and returned them to the revenue cutter. The Society's volunteers, the second crew, had rescued six more and had them over to the *Dexter* where they were taken aboard.

Meanwhile from the *Dexter* came another small boat to help. Four volunteers rowed the gig with Lieutenant Charles Kennedy in charge. They moved the gig close to the foremast and shouted to those in the rigging to jump. Up in the rigging were two teenage brothers, another passenger, a member of the crew of the *Columbus* and her Captain, Schuyler E. Wright. There were two other figures in the rigging, both of them frozen and stiff.

Lt. John U. Rhodes took a small boat with a volunteer crew on two trips to the *City of Columbus*, returning to the *Dexter* with nine frozen survivors from the rigging. Rhodes swam toward the wrecked vessel on a third try and was injured by floating debris.

The first to jump was the seaman John White of Prince Edward Island. When he surfaced the men in the gig moved in and pulled him out of the water. Next came the boys-George W. Farnsworth, 14, of Townsend, Massachusetts and his brother Hubbard M.,who was 17. They held hands, looked at each other, and leaped together.They swam to the *Dexter's* gig and were lifted aboard. The last passenger was John L. Cook of Portland, Maine who plunged, and swam to the rescue boat. Captain Wright, although unable to swim, jumped and was thrashing in the water when the gig came alongside; he was rescued with boathooks. Later Cook claimed he was the last to leave the stricken ship and not the Captain. At the official inquiry into the disaster, Wright said he was the last survivor.

In the final rescue efforts John Rhodes went once again to the water. He made two trips to reach the *City of Columbus*. On the first one he tried swimming with a rope line, but he was thwarted by floating wreckage that injured his leg, forcing him to turn back.

After his cuts were treated on the *Dexter,* Rhodes and two seamen made their way in a dinghy to the lee side of the *Columbus* where they tied up to mainmast stays. Then Rhodes and a seaman climbed into the shrouds, hauling a rope. There they reached Caleb Richardson, a 52-year old Boston businessman, who had already that day lost his wife some 12 hours before. Richardson was so frozen and tangled in the ratlines that Rhodes put a bowline under his arms and across his chest and then cut the lines to free him. As the two rescuers made ready to lower him, Richardson suddenly raised both arms, the rope flew over his head and he dropped like a stone to the water. There he was recovered and hooked into the dinghy.

The lieutenant and the seaman quickly climbed back up in the shrouds where the last person appeared stiff and lifeless. When the two rescuers reached him they saw that he was breathing. So they carefully cut him loose from the ratlines and then lowered him to the dinghy.

The four started back for the *Samuel Dexter* and its warm interior. The two would be safe there and could recover. The small boat moved past the submerged hull and the second mast toward the stern with the last two survivors. They rowed clear of the sunken hull and turned the bow of the dinghy northward with a short distance to go. Rhodes and the two seamen had to be feeling good about the way this final trip had been, but somewhere along the way Caleb Richardson died. So did the other man. It was after four o'clock Friday afternoon and the two had been in those frozen shrouds for at least a dozen hours, trying to stay alive against frigid winds and freezing spray. Their bodies were taken to the revenue cutter.

Late that Friday afternoon with no one else to rescue, the *Samuel Dexter* steamed across the sound to New Bedford. A reporter from the local newspaper, The Morning Mercury, heard the story of the death of the *City of Columbus* and the heroic efforts of rescue from several officers of the *Dexter*. The reporter, Zephaniah W. Pease, then saw and interviewed the survivors in the wardroom of the *Dexter*. Many of them wore clothing from the crew and officers. An engineer had dressed their many cuts and bruises. Pease wrote the story for his morning edition then wired the Boston Globe with details. The wire services picked it up under cooperative arrangements with their members and then the world knew about the tragedy at Devil's Bridge. Within a day or

The Samuel Dexter is shown at left on her rescue missicn to the *City of Columbus* January 18,1884. A Gay Head Indian surfboat is ahead of the *Dexter* in the gale-swept water off Martha's Vineyard. An officer on the *Dexter* drew this scene that was redrawn by staff artists at Harper's Weekly.

The sunken wreck of the steamer *City of Columbus* off Gay Head, Martha's Vineyard. Her deck houses and stack were swept away. Two visitors appear in a rowboat near the stern mast. It was here or near by that seventy-five passengers died, as did twenty-eight of the crew.

so reporters from major newspapers and from national weeklies arrived by train at New Bedford where they hired a tug to take them to the wreck and to Gay Head.

Captain Wright sent a telegram to the owners of the Boston and Savannah Steamship Company in Boston. They hired a special train on the Old Colony Railroad that arrived early the next day, Saturday. The story of the disaster had a major impact on New England and beyond. Day after day newspapers had stories on the tragedy. But the heavy and immediate impact was on relatives and friends of the 103 victims. Those people went to the office of the ship company in Boston and took trains to New Bedford where a temporary morgue had been created in a stable and on Martha's Vineyard at several locations containing bodies that had been picked up in various places in the water and on beaches.

The president of the steamship line along with a wrecking master and an insurance adjuster arrived in New Bedford on that special train and they were taken to the wreck scene by the cutter *Dexter*. Meanwhile the 17 survivors from the *Dexter* were taken to company headquarters in Boston by that special train, from there they left for their homes; some of them accompanied by relatives.

Later on Saturday morning Indians from the shore shouted to those on the tug *Nellie* from New Bedford slowly cruising by. The Indians said there were survivors in several homes back behind the beach. The *Nellie* steamed slowly through the area and recovered five bodies wearing life preservers. The five were taken to the temporary morgue in that stable and put in pine boxes next to the four dead brought in the night before by the *Samuel Dexter*.

The papers carried the names of survivors from the *Dexter* along with the stories of the wreck and its aftermath. There were, of course, the lucky eleven on Martha's Vineyard, but their names and their number were not known, so there was still hope in the minds of the hundreds of searchers who came to New Bedford.

As those looking for loved ones slowly filed through the cold temporary morgue the top of a pine box would be lifted for them to peer inside illuminated by the pale yellow light from an oil lamp. Since there were only nine bodies at that time most of the searchers would lean over to look and then shake their heads before moving to the next box. There was an overwhelming sadness to that ghastly scene and almost no conversation.

That Saturday night no one recognized a beautiful young woman who obviously had dressed in great haste when the *City of Columbus* struck Devil's Bridge. She had put on a blouse and skirt over her nightgown. A pair of tiny shoes was held to her skirt by ice. She was 23-year old Mrs. Darius K. Small who had been visiting her in-laws at Chatham on Cape Cod. Anticipating a family reunion when she returned to her home in Beaufort, South Carolina, Mrs. Small was traveling with her two-year old son, Darius, Junior. The little boy was among the missing. In that stable, in that box one could just barely see in that dim light a cameo ring on her finger with one word engraved-"Darling."
All day Saturday small craft out of Vineyard Haven were searching for the dead in the harbor and along the coast to Devil's Bridge. They had found eleven bodies from the *Columbus* which were on the floor under a sheet in a hall in Vineyard Haven. The medical examiner had attached a number to each corpse and a description.

A group of reporters and relatives went to the Vineyard Sunday morning in foul weather on the tug *Nellie* from New Bedford. They passed the wreck of the steamer and watched in silence as huge combers struck the bow of the death ship and foamed around the masts. Up in the shrouds there were still some dark clothes and blankets frozen there. The jib, although tattered, flapped wildly in the heavy wind.

The tug lowered a small boat, carrying some reporters and a few relatives of the missing. At the lighthouse they walked in on a crew member, William C. Spaulding, the purser from Boston. He had jumped clear and was picked up by the first Gay Head Humane Society lifeboat. At the home of one of the Gay Head Indians who rowed the first lifeboat they found Fred H. Tibbetts of Somerville, Massachusetts, the only passenger among the seven rescued in that first attempt. At still another Wampanoag Indian home was the Quartermaster Roderick McDonald and the other three who were with him in the first metal lifeboat from the *Columbus*.

At a little church on a hill in Gay Head there were five bodies laid out on the floor. When a man from Lynn, Massachusetts walked in, he recognized his niece on the floor, and he knelt by her, gently cradling her head in his arms, swaying slowly and talking softly to her while he wept.

At a small house by the water and still farther along the shore there were more of the dead passengers and crew from the final trip of the *City of Columbus*. On Monday the Martha's Vineyard Medical Examiner ordered that all bodies be brought to a tug to be taken to New Bedford. For many days relatives and friends of the missing hired tugs at New Bedford for the trip to Gay Head where they went ashore in small boats to walk the beaches, searching.

That fatal trip of the *City of Columbus* produced heroes and America wanted to honor them-and it did. Representative John D. Long, of Massachusetts, introduced a resolution to thank the officers and crew of the United States Revenue Cutter, *Samuel Dexter,* and "the men who manned the lifeboats from shore." The joint resolution in Congress also singled out Lieutenant John U. Rhodes for his role in rescuing the two men in the rigging "at the imminent peril of his life."

The nation's president, Chester A. Arthur, directed that Rhodes be promoted 21 numbers in grade. A group in North Carolina ordered Rhodes a solid gold Maltese cross. Rhodes's home state of Connecticut passed a resolution honoring him for his "gallant conduct." Stories about him appeared in newspapers and magazines far and wide. The Massachusetts Humane Society gave Rhodes its highest award-a gold medal. There were silver medals for Captain Gabrielson and Lieutenant Kennedy. For the crew of the *Dexter* the Society awarded $200 to be split among them.

There were fund-raising efforts in a number of directions. The Humane Society gave each of the volunteers from the two Gay Head lifesaving crews $25 and a silver medal. The six men from Squibnocket each received $15 and a bronze medal. The five crewmen who launched the whaleboat that sank each received $10 and a certificate. The keeper of Gay Head Light, Horatio N. Pease, who first saw the *City of Columbus* and had word sent to raise crews of volunteers received $25 and a certificate. Newspapers had raised from the public $3,500 for the Gay Head Indians who assisted in the *City of Columbus* tragedy. If a person was involved in any way in helping the victims they were considered entitled to some money. In all 88 were given sums ranging from $160 for the lifeboat crew members down to $5.

The public, after reading of the exploits of Lieutenant Rhodes, contributed a total of $2,053 for him, but he insisted that it be shared with his shipmates, officers and crew. To each of the other six officers Rhodes gave $150. To the crew he gave each a new uniform, a mattress and bedding.

From the public came contributions that totaled $1,000 for the widow of Third Engineer William Murray, of Boston. The papers reported that she was living in poverty after her husband's death at Devil's Bridge.

One of the passengers who received a purse from the other survivors was Captain Thomas R. Hammond of Gouldsboro, Maine. He was presented $100 for helping them.

On the morning of February 6, 1884 the official inquiry into the loss of the *City of Columbus* opened in a building on the Boston waterfront. In charge was Andrew Burnham of the United States Steamboat Inspection Service. He was assisted by his colleague, Andrew J. Savage, both of whom were the Boston officers and inspectors of that service. Under U.S. maritime law the investigation was limited to the behavior of licensed ship's officers, those of the *City of Columbus* and, as it later turned out, the Captain and the other two from the *Glaucus*. Pressure from the newspapers had seen to that.

The first witness was Captain Schuyler E. Wright of Dartmouth, Massachusetts, master of the *City of Columbus*. Wright repeatedly said he could not explain why his ship was off its course to the southward and struck Devil's Bridge rather than passing the Vineyard and moving out to sea for its routine run to Savannah. The officer in charge of the ship, Augustus Harding, was the second mate and in command in the pilot house. Maybe Harding could have explained the tragedy had he lived through it.

Captain Wright told the inspectors at the inquiry that he, Wright, after passing Nobska Light at Woods Hole took his bearings from Tarpaulin Cove Light, about half way down Naushon Island, the largest of the Elizabeth Islands, on the opposite side of Vineyard Sound from Martha's Vineyard. Wright said that he told Harding, the second mate, "to go west southwest when Tarpaulin Cove Light bears north." If that course had been followed the *City of Columbus* would have passed Devil's Bridge in open water by at least three-quarters of a mile. It was a clear night with a moon. The lights at Nobska, Tarpaulin Cove and Gay Head clearly visible. The weather was not a problem.

In their report the inspectors wrote "When the last course was given to Mate Harding by Captain Wright, the ship was where any ordinary landsman could have taken her safely past the fatal rocks without chart or compass, if any note whatever had been taken of either lights or land."

The *City of Columbus* was making ten knots through Vineyard Sound. Captain Wright was lying down on the floor of his cabin with his back against the steam radiator and his head on his elbow. His door to the pilot house was always open and he was within a couple of feet of the quartermaster and Mate Harding. It was warm by the radiator; the Captain might have dozed off. Harding told Roderick McDonald, the quartermaster, not to steer to leeward of the course. McDonald testified at the inquiry that he didn't, although the wind was heavy on the starboard bow." Sometime later the watchman at the bow came to the pilot house and mate Harding lowered a window through which the watchman shouted "buoy on the port bow!" Harding told the quartermaster "port!"

Captain Wright leaped to his feet and shouted: "Hard aport!" He thought he would see a schooner or some other sailing craft being overtaken just ahead, a common occurrence at night for steamers which would have to change course quickly to avoid a collision. Instead in the path of the moonlight there was the buoy marking Devil's Bridge. By turning the wheel to port the rudder would turn the vessel to starboard.

Captain Wright said,"there's the buoy, I'm outside it."

Then the *City of Columbus* struck the boulders of Devil's Bridge. Wright used the engine room telegraph to stop the engine. Then he ordered "back hard!" Wright was trying to move his ship toward open water. Again, he ordered the engine reversed. The ship had reversed about her length, water poured in through the ruptured bow, the hull filled and she sank in deeper water.

The time of the accident was 3:45 in the morning. About 25 minutes later-that is about 4:10-most of the 103 people who were lost had perished from drowning or the pounding seas in that bitterly cold water of January 18, 1884.

The verdict of the Steamboat inspectors was against Captain Wright. To quote from their verdict:"The backing of the engine we think was a fatal mistake." The diver from New London, Connecticut, Thomas Scott, testified that ahead of the wreck there was 16 feet of water and a sluiceway four or five feet deep made by the ship backing off. "To get his ship out of a bad place," Captain Wright made his mistake; he backed his vessel into deeper water when he should have rammed her forward into the shallows where she would have been above the water.

The press and the public wondered how, on such a clear night, the *City of Columbus* could have wound up a sunken wreck on Devil's Bridge, Martha's Vineyard. The inspectors in Boston wrote the following: "...the immediate and direct cause was the neglect of a man or men to watch the ship's course there can be no doubt."

All right, so who was in the wheelhouse when she struck the rocks? The quartermaster told the inquiry that he steered the course given him by the captain. But the quartermaster was not a licensed officer. Harding, the second mate, was a licensed officer, but not a licensed pilot. He was the top man standing, while the captain was lying down and not watching the shore with its lights and its familiar shapes. So Captain Wright was the licensed pilot on duty. His post was in the wheelhouse. He wasn't there.

Burnham and Savage, the two inspectors, wrote "Captain S. E. Wright was the only legal pilot on duty ...and for two hours immediately preceding and for at least one hour before she struck he was not at the point of duty of a pilot, attending to the actual duties, and, as master, for illegally delegating the performance of the duties of pilot, his license as master and pilot is hereby revoked." As a ship's officer, his career on the water was through. It has been reported that he became a watchman on a Boston pier. George A. Hough, Jr., author of the fine work, "Disaster on Devil's Bridge", wrote, "Captain Wright never made excuses, and he never explained. So he remains an enigma..."

Some major figures in New England shipping circles added their voices to those of newspapers who harshly criticized Captain Maynard Bearse and his two officers who looked at the wreck of the *City of Columbus* and sailed on to Boston. Bearse was ordered to the official inquiry where he testified that he talked over what they had seen and concluded that since no one could be seen on the wrecked vessel that they were dead or rescued and there was nothing that could be done.

Bearse admitted that "I see a good many things now I didn't see then. If the case were repeated, I would doubtless do differently." The inspectors felt from the testimony that they couldn't revoke or suspend the licenses of Bearse or his two officers.

One Boston newspaper called the behavior of the three "cruel neglect to make no proper effort to discover whether the wreck was tenanted." One of Boston's best known merchant shipowners, Robert Bennett Forbes, wrote "the verdict (on the *Glaucus* officers) is lame and impotent in the extreme."

The letter by Forbes was published in the Nautical Gazette, a weekly publication with wide distribution in maritime circles, both deepwater and coastal.

Law suits filed against the steamship company for loss of life and for suffering and exposure by survivors totaled $641,254. The Boston Marine Insurance Co. filed for damages for the $5,703 it paid out for coverage it had carried on cargo that was lost. With several other smaller claims the total asked for the loss of the *City of Columbus* was $647,652. Those suits had begun to be filed with the United States Circuit Court in Boston within a month after the disaster.

It was 15 months later when the claimants asked the court to direct the steamship company to pay its insurance money on the *Columbus* to the court itself from where it could be distributed. The court ruled that the company had a limited liability and refused the claimants' request. In its mortgage on the steamer the Boston and Savannah Line agreed to insure the vessel for at least the total of the indebtedness and its interest. Thus the insurance underwriters, American and foreign, paid the company the $416,000 still owed the banks plus interest. The court granted the paltry sum of $3,885.83 to the claimants who had filed for a total of $647,642. The stockholders had to be pleased beyond measure. Not so the survivors and relatives of the 103 who died at Devil's Bridge because of failure of command in the pilot house of the *City of Columbus*.

The wreck of the death ship was sold at auction for $660 to a towing firm in Boston. Divers examined the ship thoroughly and decided that the hull was not worth raising, but the boilers were, all four of them. They were sold to a company that installed them in a coal vessel. The compound engine was damaged and was destroyed as machinery by explosives with the parts raised for scrap metal. Some of the iron deck plates were taken to a pier in Woods Hole where souvenir hunters showed up with hammers and proceeded to remove bolts and take them home. "Look at this, a piece of the *City of Columbus*. You remember her."

Twenty-five women passengers died either in their staterooms or on deck, so did eight children. The passenger list was inaccurate, so the loss of babies was never a positive number, since the parents were not charged for their trip. However, some of them were counted. In Bay View Cemetery in South Portland, Maine there is a gravestone for Edwin F. Hutchinson and his wife, Alice E., and for their daughter Alice Frances, whose life ended when she was 17 months old.

On January 31, 1884, thirteen days after the wreck at Devil's Bridge the body of Mrs. Darius K. Small, known as Lillie, arrived at Savannah from New York on a Mallory Line ship. She was the lovely young woman - at first unidentified in the stable used as a

Above:Divers from New London surveyed the wreck. Only the boilers were salvaged.

Right: Peter Johnson, a Gay Head Indian, was a volunteer in the Humane Society's second surfboat that rescued thirteen people. Here he is posed at Austin & Stone's Dime Museum in Boston where he was on display and was cheated out of most of the money provided for him by the owner.

morgue in New Bedford - with the tiny pair of shoes frozen to her skirt. It was the end of her southern trip that began the afternoon of January 17 with her two-year old son, Darius Junior, on the *City of Columbus*. He was never found, so she was returned alone.

Miss Elizabeth R. Beach, a young woman passenger from Mansfield, Connecticut died on that fateful voyage. When her body was recovered from the ocean she was wearing just a nightgown. At her funeral in her hometown relatives and friends lifted flowers to one side and read what had been inscribed on top of her coffin: "And the sea gave up its dead."

After the wreck of a fine steamer with the hideous loss of life and all that extraordinary heroism of the lifesavers people wondered how was it possible that such an unnecessary tragedy could happen?

George A. Hough, Jr., a Falmouth newspaper editor/publisher and the author of "Disaster on Devil's Bridge", the outstanding work on this story, asked a shipmaster with many years of sea command. Hough wrote, "His is, perhaps, the most likely explanation of all - as it is the simplest."

"The steamer went to her death because she was on an ordinary voyage on an ordinary night. The ship was sound. The course was familiar. The weather was clear and offered no threat.

"Monotony of the commonplace bore heavily upon the men in the pilot house of the *Columbus*. Security, the dullness of routine, lulled them into inattention. There was nothing to keep them on their mettle.

"It is in such circumstances many accidents occur at sea," this veteran commander observes. "It is when seamen aren't expecting trouble they often run into it".

Yes.

LOSS OF THE *JASON*

One of the worst disasters to occur along the Great Beach of Cape Cod happened in the evening of December 5, 1893 at Pamet River in Truro.

The British full-rigged sailing vessel, *Jason*, of 1,512 tons stranded on the bars within sight of the shore. Of the twenty-five member crew, including deck officers, only one man survived.

The *Jason* sailed from Calcutta, India the previous February 15 with 10,000 bales of jute, a South-Asian plant whose strong fiber is used for making burlap and rope. The vessel was delayed for months undergoing repair after she was badly damaged in an Indian Ocean storm.

The *Jason* struck not far from shore on the Great Beach at South Truro about 7:30 in the evening. The sole survivor later told the lifesavers that the vessel broke in two right after she struck.

The *Jason* was first reported by the day patrol at Nauset Station shortly before 5 p.m. It had been raining earlier which soon turned to sleet and snow. Her captain was too close to the Cape and sailing north into the full and dangerous fury of a nor'easter. A

The full-rigged British ship, the *Jason*, bound for Boston had begun a troubled voyage from India, barely surviving a mighty storm in the Indian Ocean. Several months later she did not survive a brutal northeast storm that drove her to strand off South Truro, Cape Cod, December 5, 1893. The surfmen from Pamet River rowed to the wreck, looking for bodies. They found most of them to be buried in Wellfleet Cemetery.

circumstance that had so often forced a ship ashore to strand and break up in the high surf. Most sailing vessels caught in that fashion did not work their way out to sea to avoid grounding on the Cape. The captain of the *Jason* managed to keep his ship clear of the breakers past North Eastham and Wellfleet. The keeper at the Nauset Station telephoned the stations to the north about the plight of the *Jason* and those lifesavers were out at the edge of the water, watching for her.

At Cahoons Hollow, Pamet River and the Highlands stations the horses were hitched onto the beach carts, which were brought out of the boatrooms and made ready to leave quickly. It was by then such a horrid night on Cape Cod the surfmen on beach patrol had great difficulty covering their walks.

For several hours the *Jason* was invisible from the shore in that snow. The lifesavers were thinking-or at least hoping-that she had worked her way offshore and away from danger. This was wishful thinking. It did not last.

Everything changed at 7:30 when surfman Honey of the Pamet River Station ran into his station, shouting, "Hopkins (on the north patrol) has burned his signal!" Soon after Hopkins himself ran into the building and yelled,"the *Jason* hit the bars north of us, about a half mile!"

The surfmen at the station were ready and immediately headed north with the beach cart. When they arrived opposite the wreck, the shore was piled with wreckage.

There were soon sixteen lifesavers on the beach being the entire crew from both the Pamet River Station and the Highlands Station to the north.

Out on the *Jason* the boats were stowed on the bridge and Captain McMillan gave the order to clear them away. The towering surf breaking over the stricken ship smashed the boats and drove the crew into the mizzen, which soon fell and threw the sailors into that wild water. The captain and twenty-three others drowned.

The only survivor, Samuel J. Evans, put on a life-preserver and climbed into the rigging. Evans wasn't sure how he got into the water, but he came ashore clinging to a bale of jute. He staggered to his feet and walked unsteadily toward Keeper Rich of the Pamet River station who was carrying a lantern which Evans, the ship's apprentice, had seen.

A shot was fired quickly with the beach apparatus that fell across the wreck, but there was no one alive to respond. Usually a fire was built on the beach in case there were survivors, who then would know help was near. However, the gale-force winds extinguished the flames every time the fire was lit. In fact the wind was so strong the lantern on the beach cart was reduced to ground glass by the wind-driven sand. With four men added from Cahoons Hollow Station to those on the beach the search patrol totaled twenty, a number which did not include two local men who went to the beach as volunteers.

The bodies of nineteen of the victims were recovered from the surf and one caught in the rigging of the after part of the *Jason.* They were buried in Wellfleet Cemetery. The other four were never found.

Young Evans stayed several days with the lifesavers at Pamet River Station from where he traveled to Wellfleet to identify his shipmates' bodies. Keeper Rich got Evans a train ticket to Boston where he reported to the British Consul who arranged transportation to his home in England.

On January 22, William Evans, the father of Samuel, wrote a letter to the Life-Saving Service expressing "Heartfelt gratitude to Captain Rich and all of the gallant crew of the Pamet River Life-Saving Station for their successful endeavors to save the life of our beloved son."

Apprentice seaman Samuel J. Evans was the one person who survived the wreck of the *Jason*. He was discovered near a bale of jute, the ship's cargo, on the beach by several of the sixteen surfmen from Pamet River and Highlands Life-Saving Stations. Young Evans said the *Jason* broke in two on impact with the bar.

THE *CALVIN B. ORCUTT*

When there were a number of shipwrecks in an area where there was no lifesaving station, the Service responded by adding a station. Sometimes it meant moving a station such as the Davis Neck Station at Annisquam out of Ipswich Bay to Straitsmouth at Rockport, facing the Atlantic. It also meant adding stations at Nantucket, Martha's Vineyard, Knobbs Beach, Plum Island, City Point and Monomoy Point. Through the years there were others, always adding, always expanding the Service. When there was a need, there usually followed a new station.

In one case a lifesaving station was built and staffed because of one wreck. It happened on Cape Cod, on the long arm of the Great Beach.

Benjamin C. Sparrow, of South Orleans, was the district Superintendent of the Massachusetts coast. He attended many of the rescues and during the bitter weather on the five mile walk to reach the *Calvin B. Orcutt* his eyesight was severely damaged.

It was two days before Christmas in 1896 and the captain of a four-masted schooner, the *Calvin B. Orcutt*, was trying to make port for Christmas. She was headed to Norfolk, Virginia from Portland, Maine. The vessel was caught off Chatham in a blizzard with a heavy sea running. The schooner got too close to the beach. The captain ordered both anchors dropped, but the vessel dragged and she stranded on the outer beach.

Across Chatham Harbor from the *Orcutt* was the Chatham Station where the lifesavers knew of the disaster facing the crew of the stricken vessel. However, they could not launch a surf boat in those seas and their feelings were those of helplessness. A few miles north the Orleans Station was located in a better position to help the crew of the *Orcutt.* Because of the heavy snowstorm, the beach patrol from Orleans hadn't seen the schooner aground. To further complicate a rescue, the government telephone lines between stations from Chatham to Orleans were down.

There was one way to notify the Orleans Station that there was a ship ashore: the telegraph. So a surfman from the Chatham Station carried a message to the telegraph office where a wire was sent to Highland Lighthouse where there was a telegraph key used to send ship sightings to Boston. In turn from Highland light a telegram was sent to the Orleans telegraph office in the railroad station to be delivered to the Orleans Life-Saving Station. It was an emergency message and a local man, Henry Cummings, volunteered to carry the wire to the Life-Saving District Superintendent, Benjamin Sparrow. Cummings had a difficult walk through the deep snow to reach Sparrow's home where the phone lines were operating. Sparrow telephoned the Orleans Life-Saving Station at 11 o'clock that night with the word that there was a ship ashore on the outer beach.

Sparrow set out through the deep snow and bitterly cold wind to reach the wreck scene himself. It was five miles from his home. In the meantime the surfmen at Orleans were headed through the same conditions, dragging the beach cart with the Lyle Gun and a surfboat to a point opposite the stranded *Orcutt* which was rapidly being torn apart by mountainous waves.

The weather was so thick with snow that the surfmen opposite the wreck could barely see the *Calvin B. Orcutt* and when they could it was only during brief spells of lighter snow. They were frustrated and helpless, as the wreck was too far from the shore to fire a line for the breeches buoy. The waves crashing on the beach were so high that

The Old Harbor Station at Orleans on the outer beach was built and staffed in the late 1890s by the U.S. Life-Saving Service, because of the *Orcutt* disaster. It was modern, spacious and comfortable. Long after it was closed, the Cape Cod National Seashore moved it to Provincetown where it is today a lifesaving museum.

the surfboat could not be launched. In that bitter weather, they could do nothing. The nine-man crew of the *Orcutt* died and afterward it was stated "it was the only wreck on Chatham shores within memory, where men on shore, knowing of the wreck, were unable to recover the bodies dead or alive."

Pieces of the schooner were scattered for miles the next day. Superintendent Sparrow permanently damaged his eyes in that weather and many surfmen suffered from frostbite.

The Service decided to build a new station on the beach about midway between the Chatham and Orleans Stations. It was called Old Harbor and placed on the ocean opposite the town of Chatham in 1897.

The Old Harbor Life-Saving Station was moved to Provincetown on a barge by the Cape Cod National Seashore. It is a museum today.

THE FAMOUS *PORTLAND* STORM

From Saturday night, November 26, 1898 through much of the following morning of the 27th, Sunday, the winds were so strong - recorded at Highland Light in North Truro at 90 miles an hour - that surfmen from the Race Point Life-Saving Station could not walk the beach on their scheduled patrols.

The lifesavers knew they were in the middle of a northeast storm of towering proportions, one that would later be considered the worst maritime storm in history, according to the Smithsonian Institution.

More than 150 vessels-some experts say as many as 400-were lost from Block Island to Maine with most of them along the Massachusetts coast. This hurricane was a combination of two severe storms that met and combined their fury and stalled along the southeast coast of the Bay State. The powerful storms had barometers dropping on vessels from Long Island Sound to north of Cape Ann with captains heading for refuge at Vineyard Haven, Hyannis, Provincetown, Boston, Gloucester, Pigeon Cove and Newburyport.

The harbor at Vineyard Haven, Martha's Vineyard, looked like this on Tuesday following the violent *Portland* Storm. Twenty-nine sailing vessels were sunk, dis-masted or suffered heavy damage in that port during the three day hurricane. It took weeks to clean up that December.

The waterfront at Provincetown lost many wharves and wharf buildings. The worst of that violent storm was near and some of the damage was never replaced. The fishing industry was especially hard hit.

The center of attention then and since was the side-wheel steamer *Portland* that sailed out of Boston that Saturday night to become a Massachusetts coastal legend. It was thought for many years that she sank off Highland Light in Truro until the Historical Maritime Group of New England found wreckage twenty-five miles north they believe is that of this beautiful steamer.

"That stormy weather was threatening during the afternoon and early evening of that Saturday, the 26th, is not within dispute, for besides the warning of the weather bureau, the conditions were unmistakably proven by the flight of many vessels into port," according to the United States Life-Saving Service report on the storm.

The blowing snow turned into a blizzard that badly hampered vessels trying to navigate through a curtain of white and the horrible conditions of mountainous seas.

At Vineyard Haven some 40 vessels, most of them coastal schooners, anchored to wait out the bad weather. It came soon enough on Saturday night and by Monday night with the worst of the storm over, more than half of them were sunk or badly damaged but still afloat, or just barely.

The worst cyclonic force was moving east from the Great Lakes and the Weather Bureau was tracking its progress and issuing storm warnings. The other storm was moving up the coast from the south where it had brought snow to New York. With barometers dropping about 50 schooners and other wind ships sought safety Friday in Hyannis Harbor. In a light northwest wind about 15 of them sailed in the morning. When that wind turned around to the northeast, some of them returned, but several were never reported and later were listed as missing at sea and part of that overwhelming loss of ships and people.

At Provincetown there were between 25 and 30 vessels, mostly coastal schooners, that took refuge there the afternoon and evening of that first day of the storm of the century, Saturday, November 26. On board each of them barometers dropped and ship's captains, anticipating bad weather, headed there to "one of the safest on the Atlantic coast in northerly or easterly gales with up to ten fathoms of water and excellent holding ground." That quote from the Life-Saving Service report of June, 1899. Ten of the vessels of good size and several smaller ones were beached or sunk at their moorings, especially on the western end of the harbor inside Wood End. Most of the Provincetown wharves were destroyed.

Years after the storm a marine reporter, Isaac Morton Small, who lived next to Highland Lighthouse in North Truro, wrote in his diary for that night, the 26th, and the next day he personally recorded on weather instruments a wind velocity of 100 miles an hour off the water and over his house from the northeast. Small wrote "as far as the eye could reach on that Sunday morning not the least bit of blue water could be seen." Just white caps. The wind blowing the tops off the waves. As a marine reporter, Small would search the horizon for ships rounding the upper tip of Cape Cod and then telegraph their names to the Boston Chamber of Commerce. During that storm, he wrote, "the wind broke every window pane on the ocean side of the signal station at Highland Light."

Into that incredible fury a steamship sailed on what had been for eight years a routine run between Boston and Portland, Maine and return. The ship was the *S.S. Portland* of the Portland Steam Packet Company. She had been built by the New England Company in 1890 at Bath, Maine, "The City of Ships," on the Kennebec River about 30 miles or so by water from her namesake city. She was a pretty steamer, all white with twin stacks side by side and a large black iron walking beam on the top of the deckhouses, the beam being the standard exposed mechanism that helped turn side paddle-wheels. She had a wooden hull with huge decorated paddle-boxes that covered down to the guards those fixed-bucket side-wheels, an obsolete power arrangement that was the hallmark of ships designed for inland, protected water travel, rather than open ocean voyaging. In a violent seaway, the *Portland* was a vulnerable design and dangerously under-powered. She was 281 feet in length and weighed a hefty 2,283 tons.

As New York weather reports indicated snow, the general manager of the steamship line, John Liscomb, in Portland, telephoned Captain Hollis H. Blanchard to tell him to delay leaving Boston until 9 that night to see if the storm worsens. Blanchard was not around to receive the call but the message was taken for him. It is not known whether he received it. The *Portland* sailed at a little past 7, her regular time. Near the outer entrance to Boston Harbor the Down East steamer *Kennebec* passed the *Portland* as snow was beginning to fall. The *Kennebec* blasted her whistle several times to warn the *Portland* of the storm's intensity that was increasing off the coast. The *Kennebec* had sailed for Portland an hour earlier and her captain had decided to return to Boston as a safety measure.

As he moved up the coast an hour later and with the storm intensifying Captain Blanchard could have taken the *Portland* into Gloucester's inner harbor to wait out the storm, but he kept on course for Maine, sailing past Bass Rocks and Rockport where a keeper at Twin Lights on Thatcher Island later said he saw *Portland* at 9:30, her regular time and watched her disappear into the heavy snow then falling. An hour and a half later the captain of the schooner, *Grayling,* later recalled seeing the *Portland* off Cape Ann, rolling badly and with all lights dark. Someone on the *Grayling* burned a red flare and the *Portland* turned away to avoid collision.

Several other schooners spotted the *Portland* with the last sighting at 11:45 Saturday night. Hollis Blanchard took the *Portland* away from the coast in an attempt to ride out the storm there. He knew he couldn't turn his ship to go south and back to Gloucester or Boston, for to do so would have put her in the trough of those giant 60 foot waves where she would have capsized and died. So Blanchard held her into the wind, trying to ride out the storm.

At some point on that voyage the *Portland* died and all 176 people on board died with her or at least nearby.

She may have survived until some time Sunday morning or later, but there were no witnesses. It was 42 years afterward that a medical doctor, Thomas Harrison Eames, of Arlington, Massachusetts, wrote a scholarly account of the *Portland* saga for the New England Quarterly Review in 1940. Eames was fascinated by this major New England tragedy and had studied it thoroughly as an obsession for years. Since there was no one to tell him what the end of the *Portland* was like, Dr. Eames wrote it this way:

"It seems probable that the intense smashing she received through the night had weakened her, and finally the pounding of the sea under her guards opened her up and allowed tons of water to rush into the hull, flooding engine and boiler rooms, drowning the men working there, and depriving the ship of her power." Eames continued:"The water crashing into the helpless vessel would smash any lifeboats which may have remained, tear off doors, and burst through windows and ports, and washing helpless occupants of staterooms to death in a churning sea. Those inside were thrown into the icy water as the wooden deckhouse disintegrated, some being killed outright by falling beams, and other debris...many who had equipped themselves with lifebelts or grasped floating wreckage were benumbed by the frigid water and hammered so unmercifully by the gigantic waves that they soon died."

The contemporary wisdom continued for years about the location of the ship's remains with most of the thought by coast men believing the *Portland* went to the bottom five miles northeast of High Head Life-Saving Station. This widely-held theory was repeated endlessly for years because of reports of hearing steam whistles on Sunday and then the finding of wreckage and bodies from the *Portland* that came ashore first along the beach on the ocean side of the northern tip of the Cape, and then later when parts of the *Portland* and bodies from her were tossed by the surf onto the sands of Nauset Beach down to Orleans and Chatham.

It was believed that there were only two steamers in the vicinity of the High Head Life-Saving Station that Sunday: The *Portland* and a smaller cargo vessel, the *Pentagoet*, on her way to Rockland, Maine from New York with Christmas merchandise, much of it toys. The *Pentagoet*, described as a decrepit 34-year old ship 129 feet long built in Cleveland, Ohio during the Civil War. Her crew was estimated to be between 15 and 20. She vanished and her remains and those of her crew were never found.

The other steamer lost was the freighter *Pentagoet*, an old Civil War gunboat that also had a career as a Revenue Marine Cutter. Her crew of eighteen was bringing general freight, including Christmas toys, up the coast to stores in Maine. Her wreckage has never been found.

Surfmen started walking beach patrols from the stations all along the Cape early Sunday morning under the still difficult conditions, including wind that knocked men down and swirling sand that burned the face and eyes. Huge combers ran so far up the beach that the patrols had to walk on the dunes back of the beaches themselves to avoid the undertow from retreating waves that would have swept them out to sea. Afterward many surfmen described that storm as the absolute worst they had ever experienced.

About 7 o'clock Sunday evening John Johnson, a surfman from the Race Point Station saw an object on the beach and went over to investigate. He found a life preserver marked *"Str. Portland* of Portland." Within four hours the beach for miles was littered with wreckage from the ship with much of it near the High Head Station. There were creamery cans from the galley, life preservers, deck chairs, stateroom doors, oars from lifeboats, railings from stairs, a boat hook, a quarterboard with the name "*Portland*" from the pilot house and many pieces of shattered wood.

What the men of the Life-Saving Service feared discovering happened after three o' clock Monday morning, for it was then that the first body wearing a *Portland* life jacket washed ashore near High Head. There were others that soon appeared along the beach some wearing life jackets and others not. Some were fully dressed while others were nude, "showing that some of the passengers," in the words of Isaac Small," had evidently retired to their staterooms in the earlier hours of the voyage and were made so ill by the terrible pitching and rolling that they made no exertion to dress themselves before the ship went down." For all those aboard the *Portland*, passengers and crew, all 176 of them, it had to be a trip of terror. The estimates of the number of bodies recovered range from 28 to 60. For three weeks after that weekend storm, bodies and wreckage from the *Portland* surfaced and came ashore as far south as Chatham.

Near the Peaked Hill Bars Station three people died when the schooner *Albert L. Butler* was driven ashore about 10 o'clock on Sunday morning. The high winds and huge waves pounded the ship into helplessness off the northern tip of the Cape. The *Butler* was carrying a cargo of logwood from Jamaica to Boston when a mammoth sea covered the ship, sending the deck cargo in all directions, while smashing doors and hatches and washing away booms and gaffs and puncturing the sails. The storm pushed her into shoal water and drove her broadside toward land where the heavy sea slammed her onto the beach as a total wreck.

The Life-Saving Service discovered her when surfman B. F. Henderson on his eastern beach patrol from the Peaked Hill Bars Station was talking to his counterpart, Benjamin Kelley, who had come the other direction from High Head Station. They met at the halfway house which, although back from the beach, had the high surf breaking over it, so they did not enter the small structure for fear it might wash away from its foundation. The white crests of the breakers were so high they could see nothing beyond it toward the ocean.

The two had stopped only briefly before starting back toward their respective stations when suddenly the *Albert L. Butler* broke through the sleet, snow and spray and crashed upon the beach, startling them. The wreck was just to the northwest of the halfway house, so the responsibility for rescue fell to the Peaked Hill Bars Station. Henderson and Kelley started for their stations, fighting snow and wind-blown sand drifts with Henderson reaching Peaked Hill Station to carry the alarm in a half hour. That was considered good time given the conditions. Keeper Cook issued a burst of orders to call out his crew, clear the sand and snow from the boat-room doors, break out the light cart beach wagon and hitch up the horse. Within ten minutes they were on their way through the flying particles of sand mixed with sleet and blasted all the way by swirling snow and bitter wind. It was a desperate trip of several miles with the men pushing and pulling the cart's wheels while the horse strained and panted alongside the men. They were forced to stop several times exhausted from the difficulty of traversing foot-deep snow and loose sand all the while buffeted by the sleet and endless wind. The lifesavers had covered the distance in about an hour and they set up the beach apparatus opposite the wreck when they saw a man lying on the bank unhurt and in fair condition. So they turned their attention to rescuing the four men they could see in the rigging about 50 yards away. Not a difficult job, they thought, since they were set up with the Lyle gun on the bank about 25 feet above the deck of the schooner. The Keeper fired a No. 7 line, which landed nicely between the main and mizzen masts. But the men in the rigging made no move to get to the line, so another line was fired quickly which crossed the deck. A sailor on the schooner pulled the whip line aboard and made the tail block fast in the mizzen rigging. Then, without waiting for the hawser or the breeches buoy the sailor grabbed the whip line and headed for the shore arm over arm. Only his efforts resulted in his dropping down into the water when the line sagged. The lifesavers hauled on the whip line and pulled him through the breakers. The waves gave the sailor a tremendous pounding, but he reached shore alive and they pulled him up the bank to safety. If he had let go of the light line he would have been gone in that furious under tow. In fact that was to happen a little later with the other sailors.

A short while later the mizzenmast came down, falling forward and fouling the rigging and the whip line. Two other sailors decided to reach shore quickly, so they grabbed the whip line and headed for the shore as their companion had done. Again the line sagged and they at once were in the breakers This time the lines were snarled and could not run through the tail block. The Keeper said, "just at that moment, a big sea, I should say twenty feet high, rolled in and buried the vessel, men, and everything. I had hold of the whip line, and could feel when the sailors lost their grasp....and that was the last we knew of them." Then from right under the bank came a voice asking for help up. It was from the other sailor in the rigging who had jumped into the breakers and made it to shore, although no one had seen him reach the beach until suddenly there he was climbing the bank. A surfman reached down and helped him over the top.

The Keeper asked the two sailors from the wreck if there were any more on board the *Butler* and they insisted there were not. The one passenger, a black man from Jamaica, had fallen from the rigging while the schooner was being driven toward the Cape Cod beach by the fury of the storm. He vanished in the water.

The two rescued sailors were not in the best condition, so Keeper Cook sent them back to the Peaked Hill Bars Station under the care of two surfmen who had arrived from the High Head Station during the efforts to save the men from the wrecked schooner.

The lifesavers then picked up the Lyle gun and the rest of the gear and along with the horse headed back for Peaked Hill Station, leaving two surfmen to watch the wreck and go on board when the tide fell to see about the vessel's condition. About 2 o'clock that afternoon the surfmen were surprised to see two men suddenly appear on the deck of the *Butler* and move to the rail and jump onto the sand at ebb tide. They were a seaman and the steward and both of them had gone into the cabin when the ship hit the beach. Their shipmates did not know they were alive. The two surfmen went aboard the wreck and after looking her over went back to Peaked Hill Bars Station with the last two survivors who stayed until Tuesday along with the other two and the Captain who stayed there a few more days.

There was always close cooperation between stations, so when Kelley, who was on beach patrol, returned to High Head after he and Henderson had seen the *Albert Butler* hit the beach, the Keeper at High Head sent several of his surfmen to the wreck to assist those from the Peaked Hill Bars Station. Extra manpower was often needed.

At Peaked Hill Bars the schooner *Albert L. Butler* was driven ashore where five of her crew were rescued by the lifesavers. The other three died. Her wooden bones were on the beach for years.

The United States Life-Saving Service estimated that between 25 and 30 vessels sought refuge in Provincetown Harbor that Saturday with at least 10 of the larger ships and several of the smaller ones driven ashore by the storm. Let us quote from the Annual Report of the Service that covered that period of time: "Some foundered at their anchors, some drifted into shoal water and pounded on the bottom until water-logged, and others were driven high and dry on the land, one of the latter, a fishing boat, having been found after the storm was over on a spot which measured one hundred feet away from the high water mark. Eight or 10 wharves were destroyed and 20 buildings were blown down."

For the lifesavers the desperate part of their work began Sunday morning around 5 o'clock when Surfman Frank C. Wages from the Wood End Station was walking the

The Wood End Life-Saving station was built the year before the *Portland* Storm. Surfmen from here did incredible rescue work nearby under horrible and threatening conditions at considerable risk to their lives.

There were thirty ships lost in Provincetown Harbor one of which was the *Lester A.. Lewis* being shown to the children of this man. During the storm this schooner stranded and was badly buffeted by the hurricane winds, forcing her crew to take to the rigging for safety. In the bitter wind and snow they froze to death.

beach around Long Point one and one-half miles to the east of his station on the morning patrol with the storm roaring around him. Through the heavy sheets of wind-driven snow and sleet, Wages saw what he thought was a schooner being driven ashore in the blizzard.

So he started back for his station, running when he could through the drifts. When Wages was closer to the station he was spotted by the keeper who was up in the lookout tower. As soon as he saw Wages, Isaac G. Fisher, went down stairs and waked the rest of his crew. Wages entered the building on the run and told Fisher what he had seen. The surfboat was then run down to the inside beach, but the wind was so strong it might blow the craft off the wagon to its destruction, so the crew launched their boat into the edge of the surf and, wading along side, pushed it toward the east-northeast and directly into that brutal wind which had shifted from the east. This went on for hours and finally about 11 o'clock the surfboat and the crew had gone more than a mile when they occasionally saw through the snow and sleet two sunken schooners off shore.

Isaac Fisher ordered his men to push their surfboat toward the combers and the sunken schooners. They manned the oars and tried pulling away from the beach without success. The wind was just too strong, pushing them back time after time. The change in wind direction had thwarted their efforts; they had to change plans-and they did. The only way to reach the stricken vessels was to move the surfboat a mile and a half along the beach which was about a quarter mile north of the closest wreck. Now remember, these men had been working in the water at the edge of the surf since before dawn. They finally reached a place to windward where they could launch and pull for the nearest wreck, the schooner *Jordan L. Mott,* a 138-ton coaster, 95 feet in length that sank while anchored. Her entire hull was under water and only the after part of the deckhouse visible besides the masts and rigging. Men could be seen up high, clinging to the ratlines.

By this time it was four o'clock Sunday afternoon and Fisher and his surfmen after waiting for a slight break in the rhythm of the waves started their boat toward the wrecked schooner and after making short progress, the wind, "fairly shrieking," Fisher said, forced them back. After dragging the surfboat farther to windward, the keeper begged help from four fishermen who were in the vicinity. With two men on each oar and Fisher steering with the long oar, they managed to reach off shore just about a hundred feet, when the winds blew them back through the breakers to the beach.

Their only choice was to move still farther to windward, so they did. For the third time they launched into the surf and pulled for the *Jordan L. Mott* and managed to reach her. They rowed to the stern where they picked up one-by-one the four men who were in the rigging more dead than alive after 15 hours exposed up high to wind, sleet and snow. Up in the rigging the body of the captain's father, Charles E. Dyer, was lashed to be recovered later. The first responsibility of the lifesavers was to the captain and his three-man crew who were in bad shape from exposure. With the strong wind pushing the small boat, the trip to land was quickly accomplished by about 5 o'clock. The captain of the Mott and one of her crew were so weak they had to be carried back to the Wood End Station.

The other schooner, the two-masted *Lester A. Lewis*, of Bangor, Maine, was also a wreck. Her captain and crew of four frozen to death in the rigging before the lifesavers could reach them. The *Lewis* was a 21-year old coaster of 240 tons and 124 feet in length. She was carrying a cargo of barrels from New York for Bangor, Maine, her home port, when she ran for refuge at Provincetown Harbor. She was valued by her owners at $4,000. She was sold a week after the storm for $100. The *Jordan L. Mott* had her cargo of coal recovered by wreckers in January

On Monday, the third day of the storm, the body of Captain Dyer's father and those of the five men still in the rigging of the *Lester A. Lewis* were recovered by the Wood End surfmen and brought ashore. The rescued survivors remained at the life saving Station for three days recuperating and then were taken to the New Haven Railroad station in Provincetown where they were given tickets home.

Among the many ships that sought refuge from the threatening storm, two schooners that anchored in Menemsha Bight, an open bay on the northern side of Martha's Vineyard, were destroyed as a result of that choice of an anchorage. Devil's Bridge where the *City of Columbus* came to grief is several miles west. The Bight is a deep curve in the land off Vineyard Sound, a major maritime highway for round-the-Cape ship traffic, and offers safety and shelter in southeasterly weather, which was the wind direction when they arrived around sundown late Saturday afternoon.

At Gay Head Lighthouse on Martha's Vineyard, the only lifesaving station on the island is that building to the right. Surfmen from here and Wamponoag Indian volunteers had several days of rescue work at Menemsha Bight.

But within a few hours, or by 8 o'clock, the wind had shifted and was rapidly approaching gale force out of the east-northeast, bringing snow, sleet and rain directly into Menemsha Bight where the vessels received a horrible pounding.

The night patrolman walking east from the Gay Head Station saw a torch burning near a shoal on the western side of the Bight and by that light made out a three-masted schooner that had to have come ashore. The surfman immediately burned a Coston flare to tell those on the schooner they had been seen. That was at about 10 o'clock, and just to be sure, he burned a second flare, and then headed for the Gay Head Station as quickly as he could.

The keeper ordered the crew to break out the surfboat and its wagon and start pulling it overland. He sent one surfman to a farmhouse for a yoke of oxen, which were badly needed because of the difficult terrain.

Once they arrived opposite the schooner, which was broadside to the shore, she appeared to be close to 2,000 feet from the beach. It was about midnight when the first attempt to launch was made. The gale was blowing directly on shore, and despite a heavy effort they could not move the surfboat from close to the beach. Twice more a launch was attempted and each effort failed against the terrible force of the gale, which blew the boat back to the edge of the surf.

While several of these efforts continued, the second schooner, the *Clara Leavitt*, of Portland, Maine, broke her chains and crashed on the bar as close as 75 feet from the first schooner, the *Amelia G. Ireland*, of New York, which the lifesavers had been trying so hard to reach with their surfboat. The *Leavitt* stranded about one o'clock and with huge breakers pushed by the gale sweeping her decks, her crew quickly went to the rigging of the three masts, which fell within 20 minutes of her grounding and the destruction of her deckhouse.

Of the seven men aboard, only one survived, Philo J. Sparrow, a seaman. As his vessel was breaking up in the fury, he found and held on to a stanchion and rode before the wind and the combers, reaching the shore. His survival was considered miraculous. Sparrow had dragged himself across the beach and, unseen by anyone, found the road and walked unsteadily to the home of Charles Ryan where he was taken care of.

After it was obvious that the *Leavitt* was pounding on the bar and breaking up with so much wreckage coming ashore, the lifesaving crew from Gay Head spread out along the beach, looking for survivors. There were none.

Before 2 a.m. there was nothing left of the 24-year old *Clara Leavitt* and her cargo of molding clay. She was bound for her homeport from Perth Amboy, New Jersey and with a fair wind from the southeast entered Menemsha Bight with a falling barometer. It was to be the last place she dropped anchor. The frustrated lifesavers from Gay Head Station could do nothing for the schooner or her six crew members who died: the captain, the mate and four seamen. The report of the Life-Saving Service points out an all too common fact of the condition of many wooden coastal vessels in those years, it reads: "the character of the wreckage from the hull showed her to have been in no condition to stand heavy weather." Unfortunately, she anchored in an exposed bay in some of the heaviest weather of the nineteenth century.

With the last of the wreckage from the *Leavitt* washing ashore, the lifesavers turned once again to the *Amelia G. Ireland* which they had failed to reach by surfboat.

The Gay Head Keeper, N.C. Hayman, was not ready to give up and he sent some of his crew to bring the beach apparatus and the Lyle gun. To and from the boathouse was a six-mile round trip and the men did not reappear until dawn with the wind at hurricane strength accompanied by sleet and snow.

The weather conditions were just as miserable as they had been during the darkness, but now the schooner could be seen with sailors in the rigging and spray flying over the hull and deckhouse. The daylight made things easier and the Lyle gun was charged with six ounces of powder to throw a projectile pulling a number 7 line. The shot did not carry as far as the wrecked schooner, so the line was hauled back with the projectile.

For the second try a lighter line, a number 4, was tied to the projectile with a five ounce charge, and again it did not reach the ship and so again the lifesavers had to withdraw it from the water and prepare for another firing. After the line was wound through the pegs of the faking box, a slightly heavier cartridge was used for this, the third attempt.

Where it went couldn't be seen, but it couldn't be hauled back, so they assumed it had fouled wreckage in the water. By this time keeper Hayman felt that the wreck was too far out to reach with a line, so he decided to try the surf boat again. It was noon and his crew had been working hard for 13 hours, but despite that he felt it was no time to give up.

A number of local fisherman had gathered near the lifesavers to help if they could. These men, Gay Head Indian watermen, many of whom had been whalers were well salted and more than willing to help. It was an old tradition with these Gay Head Indians, and Hayman chose three of them to help man the surf boat. He chose Moses P. Cooper, Linus S. Jeffers and Abram F. Cooper. Each of them had volunteered for lifesaving work with the Massachusetts Humane Society and Moses P. Cooper was a veteran of the effort that saved so many at the *City of Columbus* disaster on Devil's Bridge in January, 1884.

With the three additional men and Hayman on the steering oar the surfboat was pushed off the beach into the breakers. Even with nine men on the oars it was not enough. Once again the wind slammed the boat back onto the beach. They all rested for a spell and prepared to try once more. Hayman felt that this time they would succeed. They waited for the right moment between waves and ran and pushed off. They made 20 yards from the beach-60 feet- and were hit by huge combers and that brutal wind. The surfboat again tumbled back onto the beach. For Hayman, the keeper, it was time to try something else-again.

So the Lyle gun was prepared with an 8-ounce charge and then fired. The line broke close to the projectile. On another attempt the charge was increased to 9-ounces and the line parted again. Eight ounce charges or more usually damaged the gun carriage. It was now sunset Sunday and the surfmen had been working for 18 hours without rest or food, so Hayman sent them back to their station for both. Two of the fishermen tended the bonfire on the beach as a signal to the schooner's crew.

Hayman brought his surfmen back to try a rescue at midnight with the surf boat. They worked her around to the west and launched her into the wind and snow with eight men this time, as one of the volunteers had withdrawn. He considered it too hazardous given the weather conditions. The wind had moderated somewhat but was still blowing hard.

On the first try they made it through the breakers to open water and approached the wreck off her port bow. They were temporarily frustrated by a collection of wreckage, including sails, planks and rigging. They retreated briefly but seeing an opening pulled hard and wound up next to the martingale line under the jibboom where they threw a line to the men on the *Ireland.* Six of them quickly were taken aboard the surf boat. Now it was back to the shore, leaving the seventh crew member in the rigging where he had died from exposure. His body would be recovered later to be buried in the Gay Head Cemetery. For the rescued sailors it was a difficult walk back to the lifesaving station where they spent four days recovering before being transported on the Revenue Cutter, *Samuel Dexter,* to her homeport, New Bedford, to which she had taken so many survivors from the *City of Columbus* 14 years before.

Across the Vineyard Sound from Gay Head another ship came to grief that Sunday when the steamer *Fairfax* went ashore on Sow and Pigs reef southwest of Cuttyhunk Island. The *Fairfax,* of 2550 tons, was carrying passengers and freight from Baltimore to Boston. A passing steamer, the *H. M. Whitney*, a New York-Boston vessel of the Metropolitan Line, came by and took off her passengers and brought them to New York. The captain and crew remained on board. No one was lost. Tugs and lighters from New Bedford and a steamer from Providence took off some of her cargo of cotton over several days until they were forced back to port by rough weather. Chapman Wrecking Company was hired to remove the rest of the cargo and then refloat the *Fairfax.*

On December 5 a heavy sea was breaking over her and all attempts to pull her off the reef were stopped because of the weather. Before then wreckers had taken off 500 bales of cotton, 1,500 bags of oysters and 250 barrels of sweet potatoes and brought them to New Bedford for forwarding to Boston by train. On December 13 Chapman Company told the owners of the *Fairfax* that divers had found so much damage including huge holes in her hull that she could not be salvaged.

Several other steamships had bad experiences during the *Portland* storm, but survived. The *Gate City*, sister ship of the *City of Columbus*, anchored at Vineyard Haven Harbor for refuge but lost two life boats and some sails and had cargo shifted by her heavy rolling while there. When heading south past Chatham Saturday afternoon she ran into heavy swells and grounded twice off Monomoy Point in Pollock Rip Slue. The *Gate City* arrived at her destination, Savannah, December 1.

Another steamer, the *Horatio Hall*, had sailed from Portland for New York on that Saturday and was in the same storm as the steamer *Portland.* The *Hall*, a propeller vessel with high free board, was a much safer deep-water ship that could stand the kind of buffeting that overwhelmed the side-wheeler *Portland.* Still, it was a rugged voyage. The captain of the *Hall* was a former Provincetown citizen. He was R.L. Harding and he hove to in the wildest part of that hurricane when his ship was some 30 miles northeast of Cape Cod. She was well off the coast of Cape Ann and in the same general waters as the *Portland,* and not far from where she sank. He recorded the events of that trip in the ship's log, including the blinding snow that reduced visibility to a few yards and with the 60-foot seas breaking clear over the stack at times. The *Horatio Hall* shipped quite a bit of water and her passengers and some crew were seasick, but no one was lost. Captain Harding, who had followed the sea all his adult life, said later: "I never experienced such a trip before and I don't want to again." It was reported that his ship was the only craft between Eastport, Maine and New York that faced the hurricane and came out safe and sound.

The lessons of the steamer *Portland* and the steamer *Horatio Hall* were not lost on the shipping companies. For the next steamer purchased by the Portland Steam Packet Company to replace the *Portland* was the *Governor Dingley*, a propeller vessel of 3,826 tons with a 3,000 horsepower reciprocating engine of a similar design to the *Hall* and an excellent open ocean steamer. She was built in 1899, the year after the *Portland* Storm, by the Delaware River Iron & Shipbuiding Company of Chester, Pennsylvania, and made her first trip from Boston to Portland on December 10, 1899. In 1908 she started running on the International Line from Boston to Saint John,

New Brunswick until 1918 when she was taken over by the United States Shipping Board for service to the Canadian Maritimes until 1932. As a replacement for the *Portland,* the *Governor Dingley* had a long and safe career which ended at the ship breakers in Baltimore, Maryland in 1934.

The *Portland's* sister ship, the *Bay State*, was launched in 1896 at the same yard, the New England Shipbuilding Company, of Bath, Maine. She was held at the dock in Portland the night of November 26, because of concern for the weather. She did not sail that fateful night and continued in service until wrecked on a ledge near Portland Head at Cape Elizabeth, September 24, 1916.

This white steamship, the *Bay State,* was the *Portland's* sistership. She is shown here at her Portland, Maine wharf. The general manager of the Portland Steamship Company had the *Bay State* remain in port because of the threatening weather. She continued the Boston run until 1916 when she was wrecked near Cape Elizabeth.

Most of the vessels lost during that incredible weekend were sailing vessels and not a few coal barges, including two from Baltimore owned by the huge Consolidation Coal Company which picked up railroad-delivered coal from Western Maryland and Virginia. Both vessels were under tow from the south when they either broke loose or were cut loose from their tug in the outer reaches of Boston Harbor. *Barge No. 4* was loaded to capacity when, pushed by the wind and surf, she drifted onto Toddy Rocks which are nearly a mile from shore and northwest of Hull some time between midnight and one o' clock Saturday night or Sunday morning. She carried a crew of five, two of whom survived. The *No. 4* was less than a year old, since her launching from the New England Company's yard at Bath, Maine, the same builder that created the *Portland* eight years before. The barge captain and a sailor clung to a piece of deckhouse and were washed ashore from where they walked to a nearby house for shelter and warmth. The two were forced to leave there along with the residents by the waves that began threatening the safety of the house. Keeper Joshua James of the Point Allerton Station sent someone with a wagon to bring them all to his station to wait out the storm.

The other barge, *No. 1*, drifted with the wind to the western tip of Hull where she struck. The crew from Point Allerton and a Massachusetts Humane Society crew combined efforts and rescued all five men from the wreck at Wind Mill Point. The tug *Cumberland* that was hauling barges *No. 1* and *No. 4* had anchored to ride out the storm, but the chains parted and she managed to make it safely into Boston.

The four-masted schooner *Abel E. Babcock* came to grief on Toddy Rocks not far from where barge *No. 4* stranded and broke up. The *Babcock* was a seven-year old wooden vessel from Somers Point, New Jersey weighing 812 tons and carrying coal from Philadelphia to Boston. It is assumed that she anchored in a dangerous place, but the mountainous seas pushed her toward the shoals, dragging anchors as she went. She had no chance of survival in those conditions and all eight men aboard died, including her captain for whom she was named. That event took place close to the time that barge *No. 4* pounded to pieces nearby.

Early Sunday morning the schooner *Henry R. Tilton* came ashore about one and one-half miles northwest of Point Allerton. Again a joint effort between Joshua James's crew and local Humane Society volunteers went out in a surfboat and brought all seven crew members to safety.

Several hours-maybe two or three-before daylight Sunday morning the three-masted schooner *Calvin F. Baker* from Philadelphia with coal for Boston grounded on Little Brewster Island in a heavy snow storm that cut visibility to below one hundred yards at best. The *Baker* was an older schooner built in Bath in 1873 of 567 tons and 144 feet long. The skipper was looking for Boston Light as a guide to the south channel and safety in Boston Harbor. The swirling snow was like a cloud obscuring everything including the lighthouse until they were on top of it. Boston Light, the nation's oldest, has a lantern 89-feet above the ground and with its second order Fresnel lens could be seen for 16 miles in clear weather. That night was thick with snow and the man at the wheel of the *Calvin F. Baker* never saw the rocks until his ship ran up on them.

The keeper of Boston Light and Joshua James, the keeper of the Point Allerton Life-Saving Station, had an arrangement that if a ship was in distress near the light a signal flag would be flown to alert the lifesavers.

It was after dawn Sunday when Keeper James saw the signal flag indicating a ship in distress, so James called out his crew and four volunteers from the Humane Society and they launched the Society's lifeboat which was closer. She was rowed to Hull Gut where the tug *Ariel* had been arranged to tow the surfboat out to where she could approach the wreck now fast aground and badly damaged from pounding on the rocky shore. The crew of eight scrambled into the rigging where the steward died from exposure after many hours in the bitter cold with seas breaking over the vessel. At low water the men would descend to the deck and crouch under part of the forward deck for shelter and then with the incoming tide again seek safety aloft. The first mate was carried overboard by a huge sea and vanished. His body was never found. The second mate took a line and tried to swim to the Lighthouse Island, but he died in that furious water. He, too, was lost.

The tug *Ariel* towed the society's surfboat to a point northwest of the wreck. The tug could get no closer, so the surfboat was cast loose and the heavy rowing began which brought her alongside the *Calvin F. Baker*. The five survivors and the dead steward were brought into the surfboat and then returned to the tug that was standing by to pull everyone back to land. Four days later the five were fit and able to leave by train for home.

About noon when the crew returned to the Point Allerton Station they learned that three men were marooned on Black Rock six miles south and about a quarter mile off shore in North Cohasset. The three had sought refuge on the rock after their ship had been cut loose from the salvage tug *Underwriter* and crashed on a reef and pounded to pieces.

Keeper James sent word to a man with a team of horses in the Village of Hull and when they arrived they were hitched to the carriage and its surfboat and headed for the marooned men. James had his crew plus five volunteers of the Humane Society. By 2 o'clock they reached a favorable launching site and entered the water, heading for Black Rock. When they reached their destination the heavy sea was boiling around it and they had to wait for the right lull in the water. When it finally came, they pulled quickly toward the sailors and took them aboard and then just as quickly moved off the dangerous area. Once back at Point Allerton Station, the sailors rested and recovered. After four days they left for Boston.

During that three-day storm from Saturday evening through Monday the destruction of ships recorded was the highest number of losses in New England history. The shore from Gay Head all the way to Plum Island and Salisbury was littered with the wreckage of ships and cargo, much of it was shattered vessels. The storm's violence was felt everywhere, even inland, but it was especially appalling along the water. The United States Life-Saving Service Report on the *Portland* Storm gives a sense of the power and the havoc it created:

"The storm had swept in on top of a very high tide, which caused the rushing seas to take possession of the beaches, break through the banks, flatten the sand hills, and flood the highways, marshes, and fields beyond. The seawalls were demolished, and the summer resorts...were laid in ruins...cottages being damaged and a large number completely destroyed. Some were carried from their foundations and landed long distances inshore, others were toppled over into the surf or back of the ridges, while several were blown into collision with one another. The rocks and wreckage that were driven in by the sea blocked the roads in many places and crushed the sides of dwellings. The long stretch of shore, after the storm, presented an entirely new aspect. The terrific onset of the sea had cut through or torn away the banks and strewn the beach with vast quantities of stone. As far as the eye could reach, the whole scene was one of wreck and destruction."

On Sunday morning after 8 o'clock the combination of high tide and hurricane winds drove many people from their cottages near the shore close by the Brant Rock Life-Saving Station in Marshfield. The greatest threat to these people was from the high water caused by the storm surge. There were 25 or 30 people, including children who were in peril. There were two places where they could find shelter. One was the station itself and the other was Union Chapel, a stone building at a distance of 75 yards from the

The fishing schooner *Edgar S. Foster* was anchored on Saturday off Portsmouth, New Hampshire. The storm broke her chains and she was driven across Massachusetts Bay to crash ashore Sunday morning at the height of the hurricane a half mile from Brant Rock Life-Saving station near Plymouth. The wind and those mountainous seas pushed the *Foster* so far up on the beach that her crew of eight jumped onto land and climbed that bluff to safety. Late Monday the captain returned to the wrecked *Foster*, searching for the ship's cat. The captain was rewarded for his efforts when he found the feline, named Admiral Dewey, hiding far under a bunk-terrified.

A mile or so north of the wreck of the *Edgar Foster* another fishing schooner, the *Mertis H. Perry*, was literally thrown onto the beach by the towering seas Sunday morning. Off shore five of her crew died in that violent water. Once the *Perry* was aground the other nine climbed out on her jibboom and one by one dropped onto that bank and lived.

station but inland. The water was over the beach and around the chapel and pushed by the ocean waves and wind. Gullies were created and there was danger to anyone walking through it of being swept into the sea. Here were real threats to these people suddenly caught in a hurricane next to the shore with heavy snow-almost blinding-spray and sand driven before the wind at the storm's height. To make things worse the station was undermanned.

Three of the surfmen who had been sent out on beach patrol had not been able to get back because of the destruction along the shore and the force of the storm against which it was difficult and, at times, impossible to move. That left the keeper and three surfmen to help terrified townspeople who were at high risk and close to panic. The surfmen were able to get some townspeople to the chapel and others to the station. As a place of refuge it became one of further fear as a heavy sea on the rising tide crashed open the boat-room doors and broke through the dining room windows, breaking furniture and sweeping stoves and drowning their fires, while knocking some people off their feet. Life jackets were given to the children and the women and they were taken to the second floor and the back doors were opened so the high water could rush through the building without destroying it. The frightened townspeople, who had never seen the ocean like this, must have felt they were being pursued by the sea. In fact, they were that weekend.

After an anxious wait for the tide to ebb, those people on the second floor were led across the road by the surfmen to the chapel. There was no heat in the building, so coal was brought from the station as were blankets and dry clothes.

While the surfmen from the Brant Rock Station were busy with rescuing townspeople who were their neighbors, a fishing schooner from Boston, the *Mertis H. Perry*, crashed onto the beach about two miles north of the station before noon Sunday. She had a successful trip off Cape Cod and was heading up the coast for Boston when she was caught in the storm. Both anchors were dropped, but she kept dragging them as the storm increased Saturday night. By Sunday morning when the anchor cables broke there was only one thing to do: head for the beach, and pushed by the hurricane wind she entered the breakers off shore with waves foaming over her. At that time a crew member died of exposure on the deck, and the *Perry's* captain, without a word to anyone, jumped into the boiling sea and vanished.

The combers pounding over the light, 62-ton vessel pushed her so far up on the beach that four members of her crew climbed out on the jibboom and dropped down on the bank. Following seas swung the schooner first one way and then another and suddenly the jibboom was over and above the bank again and six more desperate sailors reached land by jumping on to it. Another fisherman fell into the surf and drowned while still another died from exposure and exhaustion. Another man who got ashore was too weak to move and they left him while seeking refuge. They reached shelter in a farmhouse and they asked men there to go back to the shore for their shipmate. He was found dead on the bluff just above the wreckage of the *Mertis H. Perry*. Of the 14 men who had gone fishing on November 21, nine of them survived.

The surfman on the north patrol was forced to turn back because the beach was impassable and the huge waves had cut through the bank and created a river of sea water through the meadows. He could go no further and so kept a lookout from a promontory. He was still a mile south of the wreck of the *Perry* and never saw her. He stayed more than an hour keeping a watch toward the sea, but the morning weather was so thick with snow that it seemed useless, so he started back for the station when he came across two men with their wives and a child huddling under a bluff from the weather. He brought them blankets from a house back of the marshes. After staying with them for a while, he worked his way back to the station, joining his crew.

The south patrol ran into similar difficulties with washouts of the beach and the bank beyond, forcing him back after he had gone a mile and a half, traveling by way of a stone path, the Marshfield dike. On the way back, he discovered the dike torn in huge gaps by the thunderous waves. The beach, the bank and the marshes beyond presented obstacles he could not pass, so he turned back for the station, which he reached hours later.

About the middle of that Sunday afternoon several men from the area arrived at the Brant Rock Station with word that the fishing schooner *Perry* was ashore and there were several bodies on board. The keeper sent two of his crew to the scene where they moved the bodies out of the reach of high tide. Two fishermen from the stricken vessel spent several days recovering at the Brant Rock Station. The lifesavers had labored on shore with the storm victims from the cottages in their immediate neighborhood.

There were uncounted wrecks and missing vessels along the South Shore of Massachusetts and in its waters off the coast. The Life-Saving Service considered the destruction of the Boston pilot boat, *Columbia*, an example that stands out among the many wrecks to illustrate the titanic nature of the *Portland* storm.

The *Columbia* came ashore more than two miles south of the North Scituate Life-Saving Station sometime between mid-morning on Sunday and a little after midnight. The precise time of her destruction on the beach is not known, because of the foul weather, combining blizzard snow conditions and hurricane winds that slowed and sometimes stopped any progress by surfmen on beach patrol. Walking along the beach was precarious from the mammoth waves which flooded through stones-some of them truly good-sized far from the shore's edge. Many of the roads that ran parallel to the beach had bulkheads for their protection, but the shock of repeated pounding by the waves on such a high tide broke them apart and scattered their material along with fresh hills and gullies through the sand. The damage to cottages and other buildings gave many shore areas the appearance of a war zone.

Before the storm there were about 100 dwellings at Scituate Point and they were almost totally destroyed with only a few standing, and those badly damaged. The village afterward gave the appearance of having been stepped on by a giant. Near Scituate Light the Massachusetts Humane Society boathouse was washed off its foundation through the harbor to the south side with the surf boat found later a long distance from its wheeled carriage, which had been swept in another direction.

On Sunday morning the surfman on the south patrol, Richard Tobin, was supposed to walk the beach to the key post three miles from the station and return between 8 o'clock and noon. It was past 3:30 when he finally got back. He told about his adventures.

"When I started from the station there was much wreckage on the beach, and the seas were coming over with such force that I was washed into the pond back of the ridge. It was blowing so hard that I was obliged to kneel down at times to get my breath. It was

The Boston Harbor pilot boat, *Columbia,* was a ruggedly-built two masted schooner that was driven ashore and onto the front of this seaside cottage far up on the beach at Scituate. She was discovered Monday morning two hours after midnight by a surfman on patrol. The starboard side was stove in with the planking broken. The stern post was torn apart and both anchor chains were hanging empty near the bow. The body of her skipper was found down the beach and identified. Three other bodies were discovered not far from the *Columbia*, but they were never identified for sure as the other four in her crew.

snowing so that I could not see any distance off shore. I kept on, and finally I had to take to the fields back of the beach. Then I was able to make better progress, and at last reached the post, and then started to return. I warned a number of people in houses near by that they had better seek safety elsewhere, as the seas were breaking up against the windows. I helped two families - the women and children - to a safe place in another house and also assisted a fisherman to haul his boats away out of danger."

To return to the story of the schooner *Columbia* she was a young vessel only four years old of about 100 tons, built especially for the pilot service, which was known for the superior qualities of its vessels. Simply put, they were designed for the roughest kind of weather.

A week before the titanic storm of November 26th and the next two days, the *Columbia* sailed out of Boston into the Atlantic with a full complement of pilots all of whom had been placed on in-bound ships. The *Columbia* then was manned by the captain and four sailors who were coming back to Boston to wait for the returning pilots after which she would sail forth to contact incoming ships in need of pilotage through the harbor's tricky channels.

Nobody knew where the *Columbia* was when the hurricane burst upon the coast for none of the five men aboard survived. From the condition of the wreck it was clear that she had anchored in an attempt to ride out the tempest. When she was discovered smashed against the side of a cottage some 250 feet from the edge of the water, her anchor cables were hanging empty and the anchors missing.

Surfman Richard Tobin of the North Scituate Station had the south patrol from 8 a.m. until noon on Sunday morning, the 27th. The conditions of wind and snow were so bad he did not return to the station until three o'clock that afternoon. The wind was so strong that he had to kneel down at times in order to catch his breath. That was a frequent complaint by many surfmen on beach patrol. The heavy, driving snow reduced visibility to, at most, 300 feet, and often less than that. The snow continued until about noon on Sunday, then commenced again that afternoon. At 9:30 that morning Tobin was standing on the veranda of that cottage which was later badly smashed when the seas tossed the *Columbia* into it. So the pilot schooner had to have come ashore after that.

The man who discovered the wrecked pilot vessel was surfman John Curran who started on the south patrol from North Scituate at midnight Sunday. The wind was still heavy, the snow was still thick and the storm seemed endless to these tired lifesavers. About a quarter to 2 that morning Curran came upon the *Columbia.* He later said,"I saw the pilot boat right in line of my patrol, lying on the beach." Curran searched but found no one alive on or near the schooner.

Curran returned at once to the station and told the keeper who immediately went to the wreck with three surfmen. They looked for survivors and found none, then waited for daybreak. On their way to the wreck they found a body and moved it to higher ground. It was later identified as that of the *Columbia's* captain.

In daylight the destruction of the schooner was evident. The whole starboard side of the vessel was split open, her planking was torn badly by the rocks as she was carried to her end at the cottage where her stern post was smashed to pieces. Her mainmast had

The beach front at Nantasket was a victim of the titanic winds and violent seas of the *Portland* Storm. Early the following week these four men were visiting the scene. They may be some sort of official inspection team, but they are unidentified. Were they property owners? Insurance adjusters? Anyway, they are well dressed and appear to have money.

So many buildings at Nantasket Beach were in collision with each other. Many had to be destroyed and new structures erected.

vanished clear out of her, and her foremast was broken off to a stub. One of her crew was found dead in the hold. An earlier search had missed him.

From Martha's Vineyard to Plum Island sea walls were demolished in most places and Brant Rock, Ocean View and Duxbury Beach-all summer resorts-were mostly ruins. More than half the cottages were badly damaged, and many of them destroyed. Many jetties built to withstand the sea were torn to pieces by the violence of the storm and then scattered along the beaches.

The protected harbor of Gloucester had a number of damaged and sunken vessels as did the harbors a few miles north at Rockport and Pigeon Cove. The storm of the nineteenth century began to moderate on Monday, November 28th, and it left behind a long trail of destruction and death for which an accurate count was never made. Too many vessels simply vanished without a trace.

It was a Massachusetts event, a maritime disaster of huge proportions. Since that violent weekend it is the *Portland* Storm, named after a white steamboat that sailed into it and ever since has been a part of history and legend.

On Friday, November 25, the day before the *Portland* Storm, the fishing schooner, *Juniata*, anchored off Cohasset at night. The seas broke the chains and carried the vessel over the rocks on the shoreline and dropped her well back of the beach. These wind ships had large crews as they often used dory fishing for their catch. The *Juniata* carried a seventeen-man crew who went over the side on dry land. They stayed at a cottage not far from here.

THE MANY WRECKS OF MONOMOY

Beginning on June 30, 1876 the U.S.L.S.S. began publishing an annual report of its activities. In District 2, the Massachusetts coast, that year there were 23 marine disasters with a loss of seven vessels. Here are the highlights of the Monomoy Station.

On May 31 the schooner, *Phoenix,* stranded on Shovelful Shoal with a cargo of ice and salt. It was the inactive season, but the keeper, George W. Baker, called in the surfmen by flying a special flag and they rowed out in the surfboat and rescued the ten men in the crew. The *Phoenix* was a total loss.

More than a week later, on June 9, there were two wrecks on the shoals the same day. A south-bound cargo of ice from Bath, Maine was lost along with the schooner carrying it. Again the surfmen had to be called in from their homes and they rescued the crew of seven. The vessel and its cargo were a total loss, amounting to $8,500. The other wreck was the stranding and loss of the schooner, *Ocean Traveler*, from Rockport, Cape Ann, with granite. The six men in the crew were brought ashore in the Monomoy surfboat. The total loss was $9,000. The surfmen who responded to the call from the keeper received $3.00 for each rescue. In those early years their monthly pay was S40. for the active season, August through the next April. The District Superintendent, Benjamin C. Sparrow, of East Orleans, received an annual salary of $1,000. He served full time. His station list had grown to fourteen with the latest being Surfside on the south shore of Nantucket.

In the following year there were 21 accidents of which nine ships were a total loss. The property saved was valued at $160,000; the lost vessels and cargoes totaled $193,000. That in itself was a rare year, as the value of property saved was almost always higher than the losses.

The Life-Saving Service annual report of June 30, 1878 recorded a total of twenty accidents along the Massachusetts coast with half of the vessels a total loss. The Monomoy wreck was on March 21 when the schooner, *Highlander*, struck Shovelful Shoal, stranded and broke up in the surf. She was spotted by the beach patrol and the crew rowed out to her and rescued the six aboard and took them to the station. The ship and cargo were lost and worth $4,800.

On November 1, 1878 the schooner, *Frank Barker*, stranded briefly and lost a man overboard. He disappeared and could not be found in the tidal rips. The vessel came off later with no damage. For District 2 the total number of accidents to ships was 26 with eight a total loss for the year ending June 30, 1879. Within that time another station was added at the south end of Fourth Cliff at Scituate.

In the year following the Massachusetts coast had 25 marine accidents with nine involving loss of the vessels. For Monomoy it was a quiet year for the Life-Saving Service. The value of property saved by surfmen was $209,000. Losses in those nine incidents were $71,000. In the report of June 30, 1881 for the preceding year there were 23 maritime accidents with the loss of six ships. The value of the lost property was placed at $41,800. The vessels and cargoes saved in the other seventeen were valued at $96,000.

On October 8 of 1880 the Chatham Station was taken down and rafted up to North Beach by a local company, Oliver Eldredge. In its old position it was no longer safe with the sea encroaching so close.

A combination of government agencies saved five men from a Canadian lumber schooner stranded on Handkerchief Shoal. The Monomoy Life-Saving crew set out in their surfboat and reached the scene about the same time the Revenue Cutter, *Gallatin*, arrived. The sailors rescued by the surfmen were transferred to the cutter which took them to Vineyard Haven. Revenue cutters were cruising the coast as a regular part of their duties. They often combined efforts with the lifesavers, as did the tenders of the Lighthouse establishment when they were in the vicinity.

The Monomoy Life-Saving Station was one of the original nine built on Cape Cod when the Life-Saving Service was created in 1872. It was the southern-most station on the Cape. This station was staffed March 20, 1873. The beach patrol walked two and one-half miles north to a halfway house where they exchanged brass checks with the surfman from the Chatham Station located one and one-quarter miles south southwest of the Chatham Lighthouse. The Monomoy Station had eight surfmen that winter with the extra being the winter man. Although there was a boatroom at the station there were additional surfboats in the larger shed off to the left in this photograph.

The month of April could be a bad time for late season storms. On April 14 around midnight with a strong wind from the north, north east a fishing schooner from Providence, the *J. G, Huntington*, from Gloucester for New York with a general cargo of fish struck on the South Point of North Beach, Chatham with all hands lost in the high surf. The following morning the north patrol from Monomoy Station discovered

This is the halfway house where surfmen met and exchanged brass checks to prove the patrol had been made. This wooden shed had an iron stove to produce a toasty interior to warm frozen surfmen on cold nights. The patrol from the Monomoy Station had a two and one-half mile walk north to this shed. To the south the walk ended at a watch house at the point. During the active storm season from August 1 to May 31 surfmen walked their beaches from dusk to dawn. In stormy weather the beach patrol was also covered during daylight hours.

The first government installation on Monomoy was the lighthouse built in 1823. By 1849 the structure was in sorry shape and was rebuilt near the same spot eight miles south of Chatham using curved cast iron plates with the interior reinforced by a brick lining. This tower with its fixed white light and a red sector to the east over the shoals was two and one-quarter miles south of the life-saving station.

bodies and spars from her rigging. A boat from the schooner also drifted south to Monomoy and when all these things were found, it was known that there had been a heavy loss of life. During the night the weather was too thick to see the Point.

On November 4, 1881 five sailors were lost-that was the entire crew-of a schooner, the *Delia Hodgkins*, which foundered four miles east of the *Pollock Rip Lightship*. She was heading south with cargo from Greenland when she was struck by a heavy storm off Monomoy. She could not be seen by the beach patrol.

The months of December and January were often the stormiest, putting vessels at risk. On January 10, 1882, a valuable schooner, the *Charles Noble Simmons*, of Bath, Maine was passing between the shoals when she got out of the channel between *Pollock Rip Lightship* and *Stone Horse Lightship* and grounded on Stone Horse Shoal. The weather was heavy and she started to pound on the bottom, opening her seams. She was spotted by the man on the south patrol who went to the station with the alarm. It took him some time, because the *Simmons* was about seven miles southeast of there.

By 1881 another station had been added, making a total of fifteen. This was between Scituate and Manomet Point. It was Gurnett's about eight miles northeast of Plymouth. Later the second "t" in the spelling was dropped.

The year end reports of 1883 and '84 were quiet as to unusual happenings on Monomoy but on March 10, 1885 there occurred the first of a number of disasters to schooners with lime from Rockland and Thomaston, Maine. On that date the schooner, *Bertha J. Fellows*, from Camden, Maine with lime from Rockport, Maine was anchored off the station in a heavy west wind when she parted her chains and drifted ashore a half mile north of the station. It happened at night and the schooner was not seen and reported until 6:45 that morning by the north patrol. The schooner stranded, bilged and burned when sea water reached the lime.

Surfman Rogers reported to William Tuttle, the keeper, who ordered the dory from the west side boathouse be brought over to the ocean side. After a hard pull the crew reached the *Fellows* at eight a.m. After the schooner's crew was landed they were given shelter and food for two days. The ever-present wreckers from Chatham negotiated with the ship's captain to strip the vessel for half the proceeds.

The schooner, *Tantamount*, struck in the haze of a summer night on the edge of Shovelful Shoal after seeing Monomoy Lighthouse and thinking it was *Shovelful Lightship*. She struck but did not fill with water until 4 o'clock that morning. She was thirty years old and her hull was not strong. After filling she slowly turned on her side. Her stranding was four miles south of the Monomoy Station whose surfmen rowed down and rescued the five-man crew in their surfboat.

Later that fall on November 28, a three-masted schooner, the *Moses Webster*, of Rockland, Maine, heading north for Boston, mistook the position of *Shovelful Shoal Lightship* and stranded and bilged.

The next day the lifesavers rowed to her in the heavy seas. The *Webster* was pounding badly and it was dangerous to run the surfboat too close. So the lifesavers asked the schooner's crew to run out the main boom and have them drop into the surfboat during a period of smooth water. Four sailors, or half the crew, were picked up and taken to a nearby Gloucester fishing schooner. Then the surfmen rowed back and rescued the other four.

The *Moses Webster* was valued at $12,000. and her cargo of Cuban sugar was worth $38,000. Everything was lost, an expensive accident.

The mystery is why a cargo from Cuba to Boston was in a schooner sailing through Nantucket Sound and Pollock Rip Channel? Why not go deep water up the coast outside Nantucket and Cape Cod rather than risk the danger on the inside route?

It was about a year later when the brig, *H. C. Sibley*, of Boston was anchored at night near Handkerchief Shoal in bitterly cold weather. She parted her chains and went free only to strand and sink on the northern part of Handkerchief Shoal. The next day, December 2nd, she was seen around daylight by surfman Marshall Eldredge.

The seas were running high and the northwest wind pelted the vessel and the spray froze wherever it struck. The surfboat was launched and every drop of water that came on board was frozen instantly. The oars were covered with ice in a few minutes and the wind was so high it would blow the oars from the oarlocks repeatedly. The surfmen had to anchor twice on the way to the *Sibley* just to bail water from the surfboat. Against that wind the lifesavers had a hard pull of three and a half hours. When they finally reached the stranded ship, some of the surfmen got aboard and helped the sailors launch one of their boats, as the surfboat was too small to take the nine man crew. So two of the *Sibley's* crew got into the surfboat and the captain and the other six of his crew were in their own boat. They followed the lifesavers to Monomoy Harbor at Powder Hole from where they walked to the station.

Near the end of the active storm season an old Maine schooner from Jonesport was the last emergency for the surfmen of Monomoy. The 91-ton, *Alexandria,* was carrying coal. She was moving through the channel on a dark night at about 2:30 next to Shovelful Shoal and taking the current on the starboard bow. The 42-year old schooner went aground, her seams opened and she filled with water and sank.

It was cloudy with no fog and the man at the wheel was navigating by following the lights in the area. Wreckers reached her and negotiated to strip the schooner for 60 per cent of the proceeds.

The surfboat brought the captain and his three sailors to the station where they stayed for 24 hours. Keeper Tuttle sailed them in his catboat to Harwich and got them passes on the railroad to their homes.

Schooners and other wooden ships that old seldom survived a stranding, their hulls could not stand the impact. Most of them that old should have been taken out of service to a back channel for a final resting place. But with so many owners there always seemed to be an excuse for just one more trip. These voyages could be dangerous and sometimes fatal to their crews.

On November 11, 1887 the Monomoy lifesavers had two disasters on their hands. A small schooner with fertilizer from Boston to Woods Hole blew her sails and after anchoring was carried by the sea to Bearse's Shoal where she grounded, turned almost on her beam ends and sank. The *Joseph Lindsey* was discovered at daybreak by the surfman on the south patrol near Monomoy Point, sometimes called Point Rip. The vessel's jib was gone, her foreboom was broken and the mainmast was sprung.

The surfman reported back to the station and the crew turned out with the surfboat,where they discovered the crew of four in their own boat tied alongside. They were rolling with the waves because by that time the wind was a gale from the west. The crew was cold and exhausted, as they had been exposed to the weather since Eleven o'clock the night before.

The lifesavers took the four-member crew into the surfboat and rowed them back to the station where three of them stayed for a day after getting new clothes from the box provided by the Women's National Relief Association. The vessel was worth $2,000 and its cargo $4,000, a total loss.

The other disaster the same day happened to the schooner, *Helen Mar*, which collided with the *Pollock Rip Lightship*, then drifted onto Handkerchief Shoal and sank at midnight. While the Monomoy Lifesavers were busy at the *Lindsey,* some of the local wreckers saw the *Helen Mar* with her masts gone along with her head gear and an American flag upside down tied to a stump of one mast.

A messenger came to the station with this news and arrived after the surfmen had returned with the crew of the *Lindsey*. The Monomoy crew launched the small surfboat and spent three hours in a six-mile row against a strong gale and a rough sea.

The spread of floating wreckage, including deck beams, planks, a knee and part of her keel, was a frightening sight to the lifesavers who realized the schooner had little time to exist with that damage. It seemed doubtful she could hold together until they reached her. The wind was blowing a gale from the west and the swell was heaving. Reaching the wreck in time, they took the crew of six into the surfboat and headed for the station with a total of thirteen men in a twenty-four foot craft. With that load their boat was in no condition to encounter heavy breakers. They reached the beach without shipping too much water.

The rescued men had been in their boat since close to midnight and they were wet, cold and sleepy. Once at the station they were fed hot food, given dry clothes and a cot for a long nap. Most of the cargo was salvaged as it usually floated in the vicinity of the wreck.

The violent sea tore apart the *Helen Mar*, a 32-year old sailing vessel that should not have been coasting at her age and condition, especially in a New England winter. She died in pieces on Handkerchief Shoal under a gray sky in a surging ocean. The weather was stormy that fall and winter and another victim was a lumber schooner from Brooksville, Maine, heading to New London at night. The *Abby Wasson* was being pushed by a strong northeast gale into Pollock Rip Slew. Through the snow squalls her captain mistook *Shovelful Lightship* for *Pollock Rip Lightship*. Surfman Marshall Eldredge saw the wreck on his beach patrol. The schooner had distress signals flying from both masts. It was an hour before the surfmen could launch a boat, because directly abreast of the station the waves were too high. Tuttle, the keeper, was growing impatient and he decided to launch farther south. He ordered the boat put on a cart and pushed about a half mile where another launch was attempted and succeeded.

Just before reaching the schooner, Tuttle, at the steering oar, turned the boat so the bow was facing windward to better meet the high breakers and let the wind back the surfboat down to the *Abby Wasson*. The surfmen shouted to the frightened crew of five men to climb out onto the end of the jibboom and be ready to drop into the surfboat directly under them. There was a wait for calmer seas while the surfboat rode up the front of high waves and down the other side. After several heavy seas had broken, there is usually a space with comparatively smooth water, although brief.

The surfmen saw smoother water and quickly pulled under the bowsprit where the schoonermen dropped into the boat. Four of them did while the other fell with his legs into the water. He grabbed the gunwale and was helped into the boat.

Then Tuttle had his men pulling hard to windward and clear of the breakers. The small sail was raised and they went around the point and into Powder Hole, the harbor on the western side.

A short while later the *Abby Wasson*, that little two-master, rolled over on her side with the masts in the water and that high wind and the breakers drove her across Pollock Rip Shoal where she drifted with the current and vanished.

Often the lifesaving crew had to move the surfboat quite a long ways to be opposite a wreck before launching. The keeper leading the horse and the surfmen are wearing their cork life jackets. Most of the rescues at Monomoy were by surfboat, as the wrecks were off on the shoals-too far from the beach to fire a line and use the breeches buoy.

The waters between Boston and New York, often covered with heavy fog, were the most congested in any area of the United States. In the waters of Nantucket and Vineyard Sound, off Cape Cod and Massachusetts Bay vessels ran the risk of collision daily with schooners, fishing vessels, yachts and coastal tugs with long lines of barges strung out over a mile behind.

Within a dozen years after the United States Life-Saving Service was created in 1871 ninety percent of the freight that moved up and down the Atlantic coast went by water and ninety percent of that went by sail-a huge fleet of mostly fore and aft coastal schooners. However, steamers were being built for the trade in increasing numbers. Within twenty years, or to be precise in 1894, more steam-propelled vessels were being built than sailing vessels, although sail would still be seen until the late 1930s.

The most popular route down the New England coast or up was across Vineyard and Nantucket Sounds. The outlet going over the shoals toward New York or farther south and the opposite way heading north was the Pollock Rip Slue, a narrow channel between two shoals.

The long coastal caravans of tugs hauling schooner-barges were involved in a great number of accidents around Pollock Rip Slue in those narrow channels given the volume of marine traffic. Maine shipyards built a great many of the coal-carrying barges like this typical one at launching in the Kennebec River in Bath, "The City of Ships." She is 202-feet long and weighed one-thousand tons.

Moored at the entrance to the Slue was the old *Pollock Rip Lightship*, Number 47. Ships in a steady stream would come down over the shoals, pass out through Pollock Rip Slue and head up the back side-or ocean side-of Cape Cod for their destinations. In those days it was one of the duties of the crew to count and log the shipping that went both ways, north and south, over the shoals. The *Pollock Rip Lightship* had the record of having counted and logged more than 500 sails in twenty-four hours going and coming through the Slue.

The *Pollock Rip Lightship* carried two red lights and sounded a steam whistle and was so designated on charts at the time. More than a mile to the west and just east of southern Monomoy Island and displaying one red light and sounding a bell was the *Stone Horse Lightship*. The third lightship in that immediate area and several miles to the southwest was called *Handkerchief* with two white lights on masts and sounding a bell.

Along the channel through Nantucket Sound to the west and about mid-way between Nantucket and Martha's Vineyard was *Cross Rip Lightship* displaying one red light and sounding a bell. There were other lightships around Cape Cod, but the one that aided vessels entering or leaving Nantucket Sound was the *Vineyard Sound Lightship* southwest of Cuttyhunk Island at the end of the Elizabeth Islands. She carried two redlights and sounded a steam whistle.

In a constant struggle to maintain the channels, the United States Government dredged along those routes through the shoals.

When the U.S.L.S.S. was extended to Cape Cod from New Jersey and Long Island in 1872 the Monomoy Station was one of the nine in the original expansion. It was two and one-quarter miles north of Monomoy Lighthouse. The Chatham Station to the north was one and one-quarter miles SSW of Chatham's twin lighthouses. The beach patrol to the north was about two and one-half miles to a small halfway house where surfmen met and exchanged brass checks with surfmen from the Chatham Station. It was a way to prove the beach walk had been made. The tiny one-room wooden building was about midway between those two stations. There was no one with whom to exchange checks with on the south patrol which ended at a watch house on Monomoy Point. When the government built its telephone line in September, 1890, connecting all the stations, half-way houses and the lighthouses, there was a telephone installed at the watch house. The line ran from there north to Provincetown.

The Monomoy Station was manned March 20, 1873 with a regular complement of six surfmen with an additional winter man until May First. The first keeper or captain was George W. Baxter of West Harwich. Baxter served until 1882 when he resigned because of ill-health.

The next keeper was William Tuttle whose last report is dated May 30, 1899. Tuttle died one month later on July 1. His tenure was the longest.

On August 4th of that year the number one surfman, the legendary Marshall W. Eldredge of South Chatham, was promoted to keeper at the Monomoy Station. Eldredge served until his death on March 17, 1902 while leading his men to rescue five men from the stranded schooner-barge *Wadena* off South Monomoy Island. That grim episode is told in detail in the chapter "Twelve Men Down.' Eight miles east of Harwich is the town of Chatham. Monomoy Island extends from Chatham toward the south into Nantucket Sound a distance of nine or ten miles. In places it was two miles wide and elsewhere narrower. It is mostly beach sand with a growth of beach grass, occasionally a brush swamp, and a few meadows. It is composed of sand dunes and hollows all sizes and shapes. The sand seems to be always in motion. The sand continues to blow and shift. Today there is a sand dune, tonight a gale of wind, tomorrow some level ground. Tomorrow a piece of level ground, tonight a sand storm, tomorrow a hollow. So it is, week in and week out, month in and month out or year in and year out. The sands of Monomoy continue to blow and shift. While along the ocean side the tides and currents move the shore. The surface of Monomoy changes and so does the shore.

Monomoy Beach is bounded on the east by the Atlantic Ocean and the seas roll in from an expanse of three thousand miles and break and pound and roar upon the strand in an unbroken line. At the southern tip of Monomoy Island you can stand and see the tide rips-the broken water-that break and boil and foam. Where for so many years the lightships off in the distance would guide shipping down over the shoals. There the tide sweeps out of the sound and meets the ocean and creates tremendous tide rips at the end of Monomoy Beach. During an easterly gale at Point Rip the ocean can be seen in the raw. If you look south you can see the shoals that run all the way to Nantucket and cut down here and there by false and major channels and through them the tide rips that break and foam and are the same shoals and tide rips that frightened the crew of the *Mayflower* and forced her to turn back in 1620.

Stone Horse Lightship was closest to Monomoy Point

At *Pollock Rip Lightship* vessels turned north through the Slue.

Cross Rip Lightship was midway between the Vineyard and Nantucket. These vessels were frequently side-swiped by passing ships.

Keeper William Tuttle and his surfmen are taking the sun on a pleasant summer day after completing one of the many drills that by incessant routine kept their skills sharp. Almond Wixon is here; so is Bearse.

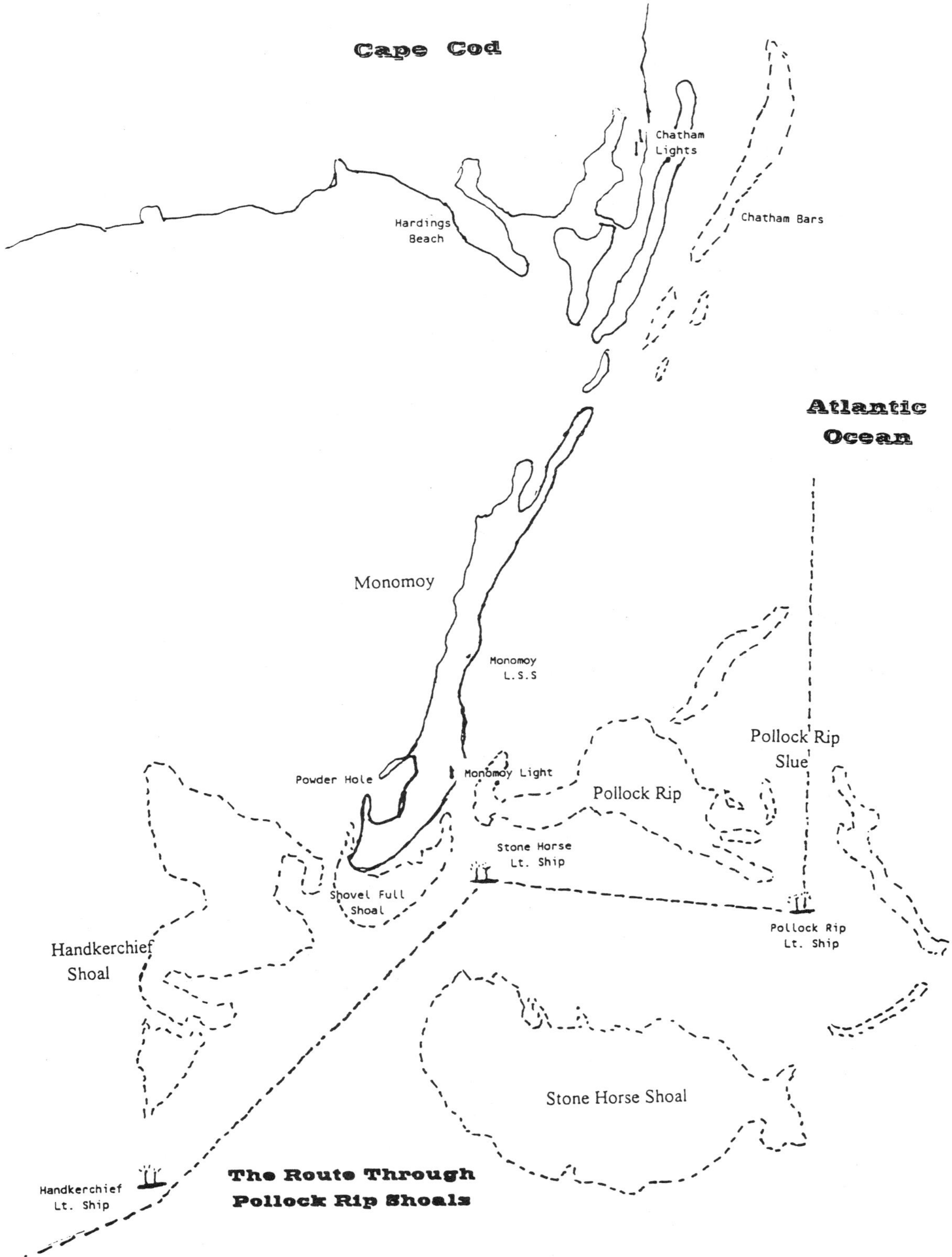
Cape Cod
Chatham Lights
Chatham Bars
Hardings Beach
Atlantic Ocean
Monomoy
Monomoy L.S.S
Pollock Rip Slue
Powder Hole
Monomoy Light
Pollock Rip
Stone Horse Lt. Ship
Shovel Full Shoal
Handkerchief Shoal
Pollock Rip Lt. Ship
Stone Horse Shoal
Handkerchief Lt. Ship
The Route Through Pollock Rip Shoals

Years ago before the Life-Saving Service was founded the ocean side of Monomoy Beach was named "The Graveyard of the Atlantic." More than any other place in the world ships pounded themselves to pieces and sailors drowned and washed up on the shore of Monomoy.

Later when word would come down that there had been a shipwreck, the keeper of the Monomoy Station would tell the surfmen starting out on beach patrol to watch for bodies in the surf.

One winter night the first watch down to Point Rip from eight o'clock until midnight was Almond Wixon of Dennisport. Half way down from the Monomoy Station to the Monomoy Lighthouse, that was built in 1823 and rebuilt twenty-six years later as a cast iron tower painted red, was a bend in the beach that was called, to nobody's surprise, "the bend."

Wixon was walking along there just back of the water when he saw something in the surf. It was moving with the waves. When a wave broke and ran up the beach, the object would roll over. When the water receded the object would roll back over and over.

Later Wixon said he thought it was a body, because he remembered that he had been told that night to look for bodies from a wreck. He was nervous, he admitted, but he waded in and kicked the object which turned out to be a bunch of seaweed all rolled up and looking like a man's body.

Wixon walked on down the beach toward Point Rip and the watch house where for a few minutes he would be out of the bitter wind. He walked along for five or ten minutes and then saw another object in the surf which rolled with the water up the beach and then back as the water would recede. This time he was not worried, saying to himself: "It's nothing. Just another bunch of seaweed."

He waded into the water again and kicked the object in front of him. As he did he had the feeling that it was similar to kicking a decayed watermelon. So he bent down and looked at the object which turned out to be a dead man rolling back and forth in the waves-not a bunch of seaweed. Almond Wixon was all by himself two miles, at least, from his life-saving station and his companions on a dark winter's night and with a dead man at his feet. He remembered the orders that whenever you found a body in the surf be sure to pull it up the beach above the high-water mark, then go for help.

Almond Wixon bent down and turned the body on its back and put his hands under the armpits and raised him up. When he did that, the dead man fell against his stomach. Wixon then walked backwards away from the water, carrying the body and half dragging it over the high-water mark and there he gently put it down in the dry sand.

He then walked and ran back to the Monomoy Station and told the keeper, William Tuttle. Immediately Tuttle called out all the surfmen in his crew and sent them both ways on the beach, looking for other bodies. Two more were found in the surf to the north, and one more to the south.

The keeper wanted to take the four bodies to the mainland to the medical examiner. So they had to be taken across Monomoy to the bay side where there was a dory. The two from the north were carried on an improvised stretcher, and the two from the south were carried on a push cart. When they were assembled, two bodies were placed in the bow of a thirteen-foot dory and the other two in the stern.

Captain Tuttle took the after thwart; surfman Bearse the forward and they began rowing for Harwichport. Once out in the bay the wind picked up from the west and they were rowing against it and making no progress. So they eased off and after two and a half hours arrived at Deep Hole Road in South Harwich. Tuttle left Bearse in charge of the bodies and the dory and headed for the undertaker's place, a man named Levi Long who was to go pick up the bodies. From there Tuttle walked to Harwich Center several miles away to the office of Dr. George Munsel, the medical examiner. After that he started walking again, this time to Harwichport where he had a short visit with his family at his home on Sea Street.

Once again Tuttle set out and from his home; he walked back to South Harwich and down to Deep Hole Road. Near the end he could see Bearse and the dory on the beach. The undertaker had already removed the bodies. While Keeper Tuttle and surfman Bearse were rowing back to Monomoy the four young men who had been found in the ocean surf were buried in simple pine boxes in the South Harwich Cemetery about ten feet in from the west fence almost in the shadow of the church. A marble marker was placed at the head of each grave. There was no service, no relatives, no friends, no flowers and no tears. The undertaker, Levi Long, read a simple prayer. That's all. They were unidentified where they died and where they are buried. Yet somewhere, someone must have known who they were, must have been waiting for them to return, and later wondered what happened. They were young men of the sea whose vessel was wrecked near Monomoy and left them to drown, and their bodies to be discovered on a lonely beach after dark and the next day buried in an unmarked grave.

Once they were out on the bay a strong westerly wind came up and helped Tuttle and Bearse row back to the western side of Monomoy. They landed about dusk that winter afternoon and pulled the dory up to the boat house.

The two set off for their station over the dunes, down the hollows and through the sand, ankle deep. They had been up-these two-since just after dawn to get an early start after most of a night searching for bodies. They had had little sleep. After that day of heavy rowing and for Tuttle miles of walking, sleep that night came early-and it came easily.

The schooner, *Royal Arch,* with coal from Port Johnson, New Jersey to Boston was lying at anchor under the lee of the beach three miles south of the Life-Saving Station, waiting out a gale. During the night the big, three master struck several times upon what seemed to be a sunken wreck. She started taking in water and her crew manned the pumps, but the water kept gaining.

The pounding started about 5 a.m. on November 29th and was discovered about 9:30 that morning by surfman Bearse on the day watch in the tower. The surfboat was launched and rowed to the scene in 45 minutes. The lifesavers took a turn on the pumps to no avail, with the water rising rapidly it became obvious she would sink.

A passing schooner, the *Henry S. Little*, stopped to help and agreed to take four crew members to Vineyard Haven along with all the baggage. The captain and the two others came ashore with the lifesavers and the following day were taken to Chatham where free passage was secured for them on the railroad.

A great deal of granite was shipped out of Vinalhaven, Maine to Boston, New York, Philadelphia, Baltimore and Washington. Specially cut blocks for public construction projects, including columns for court houses, post offices, custom buildings, and museums were a regular part of such cargo. Paving blocks for city streets were a large part of the granite quarry industry and they were loaded on schooners at the long piers.

Sailing south with such a cargo out past Matinicus Rock Light on a gray January day in 1888 was the two-masted schooner *Hunter* of Rockland, Maine. More than a day later at night the schooner was about a mile west of the *Pollock Rip Lightship* and heading for the *Stone Horse Lightship*. It was 3:30 a.m. and the weather clear when another schooner drifted under the bow of the *Hunter* and they collided. The jibboom was torn away and the vessel anchored. The other schooner left the scene. It was low tide and the sea was smooth.

At daylight on that bitterly cold morning the surfman on the morning patrol, A. F. Wixon, saw the *Hunter*, lying at anchor a mile west of *Pollock Rip Lightship*. The mainsail and foresail were set and the crew was trying to heave up the anchor. The wind had increased to a gale and the sea was rough. The *Hunter* was straining on her anchor chain and with that small crew of six it was impossible to get up the anchor, so they set a signal of distress, an upside down American flag.

The lifesavers had been watching, and as soon as the signal was put up, the surfmen launched and started pulling for the scene. In that cold wind every drop of spray that hit the surfboat turned to ice. The schooner's captain told Tuttle the vessel was deeply loaded with stone and part of her cargo was on deck and if she lay there she very likely would ice up and sink. The captain said either he leaves her there and goes ashore with the lifesavers or gets her under way and in under the lee of the beach.

Tuttle told the captain that it would work and they got underway. The lifesavers joined the other crew and together there was enough muscle to hoist the anchor. Then they sailed through the ice toward the beach. The trip was about three miles directly to windward but the current was in their favor and they made good progress and anchored opposite Monomoy Lighthouse about 11:30. The water was calm there and the *Hunter* did not ice up.

The schooner captain was pleased to be out of his dangerous position and he thanked Tuttle.

In getting along side the *Hunter* the surfboat in those long swells received several hard thumps, which started her garboards, causing her to leak badly. On their way back to shore one man had to bail constantly to stay ahead of the water.

A lumber schooner from Bangor, Maine headed for Bridgeport, Connecticut sank in a gale on the southeast part of Handkerchief Shoal August 27, 1890. The weather was thick and rainy when the wreckers at Monomoy Point saw the vessel, the *Annie E. Moore*, with the crew in the rigging. One man, Ned Grey, of Fall River, Massachusetts, was washed away.

A passing steamer, the *Lucy P. Miller*, saw the men in the rigging and rescued them. They were carried aboard the steamer and landed at Rockland, Maine. The schooner drifted onto the shoal where it completely broke up.

The missing sailor, Ned Grey, was picked up adrift clinging to a piece of wreckage. He later arrived at his home.

The following winter there was a three-day saga of the plight of a Canadian lumber schooner which stranded and sank on the eastern part of Handkerchief Shoal at night. The weather was thick and rainy and the vessel, *Florida,* could not be seen from the Monomoy Station and it was too early for the day patrol to be at the Point. However, the keeper of the Monomoy Lighthouse did see her and he walked over to the station and reported it.

The surfboat was launched into the south wind and after a hard pull, they reached the wreck.

Tuttle kept his crew on board the schooner until high tide and they tried to get her off by hoisting the head sails and forcing her bow to leeward into deeper water. Shortly after dark the pounding hull started leaking and soon she was half full of water. Toward morning the weather moderated but was still threatening and Tuttle thought he might have to take the crew to shore, so he took their personal effects and luggage back to the station. At 11 a.m. that morning they boarded her once more this time with telegrams for the captain. Two steam tugs were standing by with lines on board.

By mid-afternoon the wind came fresh from the east-north-east, so Tuttle and his men rowed back and went on board again. The tugs had left and the crew was still aboard.

The next day, the crew left the schooner in their own boat and spent the night with the wreckers at Monomoy Point and then came to the station. Tuttle and his men did not go aboard the *Florida* again as she was in charge of the wreckers.

The wreckers had thrown off her deckload of lumber to lighten her and the next night she drifted across the bay and grounded on the bar off Harwichport.

The following morning the tugs got her off.

A three-masted schooner from Camden, New Jersey sprang a leak in a heavy northeast gale and sank in mid-channel.

The accident was discovered by Marshall Eldredge, surfman on the morning patrol. The vessel, *Edith T. Gandy*, was carrying a cargo of paving stones from Lanesville, Massachusetts on Ipswich Bay to New Bedford. Lanesville was a busy shipping point for the many granite quarries in the hills back of the town.

Upon arriving Tuttle found the crew on the mast head wrapped up in the gaff topsail. A telegram to Woods Hole for a tug to come rescue the crew did not bring such a vessel, but a steam yacht, the *Wild Duck*, did respond, arriving at 4 a.m. The shipwrecked crew were put on the steam yacht where they were cared for, and otherwise made as comfortable as possible on the trip through Nantucket Sound to Woods Hole.

A schooner-barge from Philadelphia was being towed along with another barge when they were not given enough room to pass by another tow heading south in Pollock Rip Slue. The barge *Beechwood,* loaded with coal, struck the sunken wreck of the big, steam yacht *Alva*, a victim of a previous collision.

The *Beechwood's* tug ordered the barges to anchor. The collision happened at 4 a.m. on December third. During the anchoring the hawser had some slack and it caught around the propeller of the tug, which was helpless so her captain anchored.

At daylight the tug hoisted a signal of distress and Tuttle launched the surfboat.

While the lifesavers were there, the engineer managed to unsnarl the hawser from the prop and move the *Beechwood* close to shore into calm water where a suction hose could be put onto the barge to pump her out. The water was soon lowered so she could be towed to Vineyard Haven for repairs.

The surfmen helped in getting the tug alongside the barge and in handling the hose pipe.

Tuttle and his men stayed on the tug and the barge until the danger of sinking was over and then they returned to the station. The danger the lifesavers faced was recognized and the huge savings in property they made-almost 350-thousand dollars in 1892 in Massachusetts alone-finally got them a pay raise to $65 a month for the ten months of the active season. Superintendent Summer Kimball had lobbied hard for the increase.

In the middle of a brutally cold winter, lifesaving was a dangerous and chilling business. On February 7, 1895 a Canadian lumber schooner from St. John, New Brunswick was caught in the ice and carried upon Handkerchief Shoal. She had anchored near Bishop & Clerk's Lighthouse off Hyannis Harbor and became frozen in as the whole of Nantucket Sound was filled with ice. On the morning of the 9th she started to drag her anchors. No assistance could be given by the lifesavers because of the ice, so the keeper telegraphed the Revenue patrol, asking the cutter *Dexter* be sent. At daylight on the morning of the 10th the schooner, the *Prudent*, moved by the ice floes was ashore on Handkerchief Shoal with a signal of distress flying.

The surfmen launched on the morning of the 10th and after rowing through two narrow strips of ice got alongside the vessel, and found her to be in fair condition. She had lost two anchors, but their best one was left. Soon after the surfmen reached the schooner, a tug came by and agreed to tow the vessel to Vineyard Haven for $300.

The lifesavers helped raise the anchor and getting the tug's hawser fast to the *Prudent*.

The Reading Railroad had fifteen of these coastal tugs which had two crews for continuous operation, as the economy in towing was in the steady movement of ships.

Schooner-barges had a number of wrecks, mostly because of the length of the tows. Sometimes it was the tug's fault as happened April 11, 1895 when the barge *Onconta* of the Delaware & Hudson Coal Company in Pollock Rip Channel was in a collision with a schooner.

The tug attempted to haul its barges across the bow of the schooner too close and the impact damaged the barge and she started to sink. The *Onconta* swung around alongside of the *Isaac H. Tillyer* whose crew threw a line on the barge. Four of the crew climbed down on the deck of the schooner. The barge captain was the last to leave and he grabbed the line and seemed unable to climb. He slipped down between the two ships into the water and passed out of sight.

Later the tug set a distress signal and steamed in toward the station. Tuttle and his men launched immediately and met the tug, which had a message for the company agent in Boston. The lifesavers took the message to the station and called the telegraph office in Chatham.

In his accident report Tuttle blamed the officers on the tug for pulling the tow across the bow of a schooner as steamers are supposed to give sailing vessels the right of way. The six-year old schooner-barge was worth probably $25,000 and her cargo of coal $2,500. The barge captain died in the mishap.

Coastal caravans of steam tugs and barges were growing in numbers, as they offered an economical way to move coal. Of all the vessels sailing those congested waters, the tows were the least popular among other mariners.

The captain of a Canadian schooner from Prince Edward Island for Boston made an error in reckoning during thick weather and mistook the Chatham Lights for the twin lights on Thacher's Island off Rockport. The vessel, the *Nantasket*, struck on Bearse's Shoal, stranded, sprang a leak, filled with water and capsized. It was December 6, 1895.

Surfman Chase on the morning patrol saw the schooner at daylight and telephoned the station from the watch house at Monomoy Point. The lifesavers launched and rowed to the wreck and found nobody on board. So they returned to the station where, just after landing, they noticed a boat near the wreck. Assuming it to be the crew, the surfmen launched again and rowed to the schooner. The other boat contained the crew who had been aboard another schooner a mile away. The surfmen took the baggage and some of the crew while the others followed in their own boat.

The schooner's crew stayed two days at the station and then were brought to Chatham for free passes on the railroad.

The *Nantasket* was later towed to Hyannis, but never reached there as she went ashore on rocks and broke up.

Lifesavers knew a great deal about handling all kinds of vessels, since most of them were mariners for many years before joining the service. A Canadian schooner, the *Arthur M. Gibson*, needed that kind of help when, on May 22, 1896, stranded on Handkerchief Shoal in thick fog missed seeing a lightship when passing.

She was carrying coal from New York to Halifax when she stranded at 11 a.m. She was not seen until the next day, because of the continuous fog. She was finally spotted six miles southwest of the station by temporary surfman Foye on the south patrol at 4:30 a.m. The lifesavers responded with the keeper's sailboat and dory, arriving at the scene at 6:15 a.m.

The lifesaving crew rendered assistance in handling sails and anchors, then they sounded the depth of water in different directions around the schooner and when she floated, they piloted her out over the shoal. Some of these surfmen had gone to sea when they were as young as nine or ten.

The lifesavers of Monomoy had an intimate knowledge of the shoals and channels around that area and were called upon to help another vessel in distress on December 14, 1896.

The full complement of Monomoy Station surfmen in front of their beach apparatus. The number one surfman, Marshall Eldredge, is third from the right. He would become the keeper when William Tuttle left the post. The Government telephone line is strung on those short poles. The land is typical of south Monomoy.

It was 2 o'clock in the afternoon when the *Cora C. Meader* of Boothbay Harbor, Maine was heading west through what was known as "Lighthouse Channel." That route in recent years had become narrower and more crooked. Her captain was not aware of the change as he had been accustomed to going through and thought he could again. The tide was low at the time and with a loaded vessel there would have hardly been enough water in the deepest channel.

The *Meader* was a two-masted schooner with a heavy deck load of Maine lumber from Bangor for New York. As soon as she struck bottom, the captain raised a distress signal, the American flag upside down. Surfman Norwood on the south patrol spotted it at once and telephoned the station from the watch house. The crew launched the surfboat and reached the *Meader* an hour later.

The lifesavers gave assistance in handling the sails and the anchor, and when the *Cora C. Meader* floated with the tide, they piloted her out through the channel. These examples are only a sample of the emergencies faced by the surfmen of Monomoy and elsewhere on the Massachusetts Coast. The first obligation for the men of the Life-Saving Service was to save lives. The second was to save property.

On August 16, 1897 a big, four-masted schooner from Philadelphia, the *Isaiah Hart,* with a cargo of coal for Boston drifted out of the channel and grounded on Pollock Rip Shoal. The lifesavers rowed to her in the surfboat but could do nothing until the tide and current changed. The surfmen waited on board and hailed the U.S. Lighthouse tender *Azalea* which moved in and took a hawser. After awhile it broke. The tide turned to flood and the schooner moved around and floated.

The lifesavers helped handle anchors and sails. Lighthouse tenders and Revenue cutters helped in emergencies, too.

A problem that bothered the keeper at Monomoy was the distance his station was from Monomoy Point. In the middle of February, 1898, William Tuttle finally included the idea of another station farther south.

The schooner-barge *Excelsior* of the Reading Railroad Company of Philadelphia and another barge, the *St. Nicholas*, were anchored about eight and one-half miles west by south of the Monomoy Station. The *Excelsior* broke her chains, drifted ashore and sank. In spite of the northwest gale and swirling snow, the crew of the *Handkerchief Lightship* saw the barge stranded on the shoals.

The following day it was clear enough for the beach patrol to report the wreck. The lifesavers launched and reached the barge, but saw no signs of life on her. The barge was soon breaking up in the heavy seas, and there was a tug in the vicinity, so the surfmen returned to the station. The next day a man's body was taken from the wreck by crew members from the *Handkerchief Lightship* and brought to the Monomoy Station. Tuttle took charge of the body and saw to it that it was brought to the medical examiner in Harwich.

Tuttle wrote in his report of the wreck and the death:"It is possible that she might have been seen from the point had there been a station there and a proper lookout maintained."

Almost two months later on March 13, a three-masted Canadian schooner ran a wrong course and stranded on the extreme southwest part of Handkerchief Shoal. The 700-ton *Avona* was carrying an $80,000 cargo of hides from Argentina to Boston on a clear morning, March 13, 1898, at 8 o'clock with all lightships and buoys in plain view. Obviously, something was wrong-very wrong.

The course steered from the *Cross Rip Lightship* midway between the Vineyard and Nantucket was more than a point off, the mate later told the lifesavers. She was spotted immediately by the morning patrol at the point who ran to the watch house and telephoned the station.

Tuttle ordered a launch of the surfboat and there was a full crew at the oars. The *Avona* was six and one-half nautical miles from the station and it took two and one-half hours to pull that distance through rough water and into a fresh south by southwest wind.

When the lifesavers boarded her they found the problem to be the officers who were intoxicated and in no condition to be navigating a ship. The surfmen helped hoist the sails and the schooner finally floated off at 1:30 that afternoon. A tug, the *Herald*, came by and bargained to tow her to Vineyard Haven. The wind was favorable for the schooner to sail to Boston, and there was no reason the vessel could not have proceeded provided her officers had been sober. However, they were not, so the *Herald* towed the *Avona* to Vineyard Haven where the ship's officers had a chance to sleep off their condition.

The cargo of hides was intended for Massachusetts shoe factories. Shipments of hides from South America were some of the most valuable cargo carried to New England.

It was the last watch of the active season on May 31, 1898 when surfman I. T. Foye on the south patrol discovered a three-masted schooner, the *Isaiah K. Stetson*, stranded on the southeast part of Shovelful Shoal. She had mistaken the light of a vessel at anchor for that of a lightship. The *Stetson* grounded at 9:30 at night and was seen by Foye fifteen minutes later.

The crew launched and pulled to the schooner in one hour. It was shortly after high tide and since nothing could be done until high water, the surfmen left and returned the next morning.

It was evident the schooner had to be lightened in order to get off. Many hands would be needed, so the captain contracted with the Monomoy wreckers to put the schooner afloat.

All the spars in her deck load were thrown over the side as was one-third of her cargo of ice in her hold. The surfmen sounded out a channel for the schooner's captain to steer away from the shoals, which he did shortly.

A sudden change of wind proved disastrous for a big barkentine in Pollock Rip Slue which could not be brought under control. The *Harriet S. Jackson* was driven out of the narrow channel and struck on Pollock Rip, sprang a leak, stranded and sank.

The accident happened late in the evening of September 20, 1898 and was seen by surfmen Chase and Crowell who sounded the alarm and helped launch the surfboat which reached the *Jackson* in fifteen minutes. When she floated after several hours it was discovered she could not be kept from sinking even with four pumps working. Her captain headed for shore where he wanted to anchor and keep his ship afloat until daylight.

The weather was so thick with fog the vessel could not see the shore and she struck without her crew seeing the beach. That's just as well, as she would have sunk before morning had she anchored and her crew of eight would have left in their own boat and possibly met with disaster as there was a rough sea running.

As to leaving in their own boat, the Life-Saving Service never recommended sailors do that as they often did not have the experience with small boats that surfmen had, and furthermore the lifesavers had seaworthy surfboats designed for such emergencies.

In May of 1899 Keeper William T. Tuttle of the Monomoy Station again made a plea for a new station to be built and manned at Monomoy Point. His complaint was with the wreckers. Tuttle wrote it this way: "It seems to be the policy of the wreckers to reach a vessel as soon as possible and conclude a bargain before we can get there. As about nine-tenths of the strandings occur near Monomoy Point the wreckers have all the advantage of distance. They can reach a vessel ahead of us every time in spite of our best endeavors. This station had always labored under a great disadvantage in that respect. The greater part of work has been away off at arms length with nearly always wreckers to compete with." His pleas would finally be answered-but not yet.

Tuttle's complaint was part of a report he filed with the District Superintendent concerning the stranding of a fishing schooner, the *Thomas Brundage* of Boston. The vessel grounded in a thick fog more than three miles southwest of the station.

The wreckers work a vessel stranded to the right background. Their horses are covered against the winter weather. These men made a good living by quickly reaching distressed ships ahead of the lifesavers and then setting a fee to save the vessel or remove the cargo. In bad weather the wreckers were always out.

The Number One Surfman, Marshall Eldredge, on the south patrol spotted the *Brundage* at 10 o'clock in the morning of May 8, 1899. The lifesavers launched and when they arrived at the stranded schooner the wreckers were already aboard and throwing off ice to lighten her. The captain had made a contract with the wreckers to float her, so there was nothing for the surfmen to do after they arrived.

That wreck report was the last one Tuttle wrote as keeper. He retired in July and died within a month at his home in Harwich. His replacement was the Number One surfman, Marshall W. Eldredge.

Those old schooners used as "limers" burned frequently all along the coast wherever they sailed. If one of them sprang a leak and the salt water got near the casks of lime, there was a fire. Some of them were in such sorry shape it was often a wonder if they could stand the strain of leaving Penobscot Bay or sailing across the Gulf of Maine and Massachusetts Bay to reach Monomoy without sinking or burning or both.

Which, of course, leads us to a story. The small, two-masted schooner *Laura Robertson* of Rockland, Maine was just such a vessel. She was thirty years old and worth, optimistically, one-thousand dollars. In other words, very little. She was seen at daylight by surfman Small on the north patrol when she was on fire abreast of the Chatham Station and drifting south.

It was April 7, 1900, a clear morning with a strong wind and a rough sea. The *Robertson* was about two miles above the Monomoy Station and about the same distance off shore. The lifesavers saw her boat leave the vessel, but they could not launch the surfboat as the breakers were too high.

The surfmen put a flag on a pole and signaled the schooner's crew rowing off shore to head south. When they reached four miles below the station, the water was smooth enough to land. The surfmen then signaled the schoonermen to pull for the beach, which they did and landed.

The five men were escorted to the station and fed and later that afternoon were sailed to Chatham and then escorted to the railroad station and secured tickets for their homes.

The report was the first wreck notice filed by the new keeper Marshall W. Eldredge of South Chatham.

That same year on September 3rd, another old "limer," the *Lizzie Smith* of Sedgwick, Maine stranded on the extreme edge of Shovelful Shoal, started leaking and soon filled with water, caught fire and burned.

It was a dark night and by 4 a.m. surfman Nickerson on beach patrol saw the flames and phoned the station from the watch house. The schooner was five miles southwest of the station and the crew rowed there in two hours.

The little 77-ton schooner only had a three-man crew and none of them were acquainted with the area or had a chart. The lifesavers took the three men into the surfboat and landed them in Chatham and secured for them railroad passes to Boston.

The *Lizzie Smith* was at least thirty years old and not in good condition.

In the year 1901 there were at least three cases of ship wrecks on the shoals near Monomoy in which, because of fog, the vessels were not seen by the lifesavers and the crews of those vessels were picked up by passing ships- schooners and steamers.

TWELVE MEN DOWN

On the night of March 11,1902 a coastal tow of a tug and two loaded barges carrying Virginia coal to Boston was slogging through Nantucket Shoals into the face of a northeast gale. The tug, *Sweepstakes*, an older vessel built in Cleveland in 1867, was 130 feet in length of 227 tons and powered by a 600 horsepower steam engine.

The barges were the *Wadena* of 1076 tons, 37 foot beam and 212 feet in length. The other was the *John C. Fitzpatrick* of 1270 tons, 39 foot beam and 242 feet long.

It was a bad night and the tug's skipper found himself making slow progress into that storm, so off the southern end of Monomoy Island he was looking for a safe anchorage to ride things out. Coastal shipping often required frequent anchoring, but in the dark the barges struck hard aground on Shovelful Shoal.

The next morning the lifesavers from the Monomoy Station rowed out to the barges and tried to float them off the shoals, which they could not do. With the wind still blowing hard out of the northeast, the Monomoy crew took the crews of five men from each of the barges and brought them to the station, arriving there at three that afternoon.

The barge owner contacted the insurance underwriters in Boston and also hired wreckers from Chatham to lighten the vessels and refloat them. The *Sweepstakes* stayed in the area for several days and then left for repairs and was replaced by another tug, the *Peter Smith*. The wreckers continued to unload the cargo of coal until the night of the 16th when the weather again turned threatening. All the men except five were taken off the *Wadena*. The five who stayed were the owner, William Mack of Cleveland; the Captain, Christian Olsen, of Boston; and three Portuguese wreckers: Walter Zeved, Robert Malanox and Manuel Enos.

The lifesavers, watching the two barges the following morning, decided the ships were in no danger, although the weather was heavy and the sea in motion. The keeper, Marshall Eldredge, and several surfmen had reached their morning decision about the barges when the phone rang-a call from the tug captain in Hyannis who inquired about the condition of the men on the *Wadena*. Eldredge thought the barge was empty of people and he became concerned by the tugman's question, and so decided to walk to the point and take a closer look.

When he reached the point after a three-mile hike he could see the *Wadena* about half a mile off shore, aground on the shoals. She appeared to be all right, but in the rigging was an American flag flying upside down-a signal of distress! He had to respond to that.

In the tiny south watch house was a telephone to the station. Eldredge called his number one surfman, Seth Ellis, told him of the distress flag and ordered Ellis to gather the others and go to the second boathouse behind the station on the western side of Monomoy, launch a surfboat and head for the point by the inside route on the bay side. In order to save time the keeper had ordered Ellis to use that inside route with the smaller of their two surfboats. The larger, safer boat was in the main boathouse above

The keeper at Monomoy, Marshall W. Eldredge, ordered his crew to row out to the *Wadena,* saying : "We must go! There is a distress flag in the rigging!"

the high water mark facing the ocean in front of the Monomoy Station. It would not be used this day.

So the six surfmen put on their storm clothes of flannel shirts, khaki pants and oil cloths and boots and struck out across the sands of Monomoy led by their number one, Seth Linwood Ellis, for the boathouse on the bay side.

The six were Valentine D. Nickerson, married with four daughters, he was hard and wiry, six feet tall and weighing 185 pounds and considered a typical Cape Cod waterman, experienced with the sea, as were the others. Arthur W. Rogers of North Harwich was married with one child. He was the same size as Nickerson. Osborne F. Chase and his wife had three daughters. Chase, like Nickerson, lived in Harwich. Edgar C. Small of Harwich Center was married with a son and a daughter. Elijah Kendrick lived with his wife, son and daughter next to the baseball field in Harwich Center. Isaac Thomas Foye of South Chatham was the sixth surfman that day. Married with no children, Foye was of average height with one distinction: he was such a good-looking man that local women referred to him as "the handsome surfman."

Seth Ellis was a rugged waterman six feet tall and weighed 190 pounds, who lived in Harwichport with his wife and son. He was 43 years old and had been around the water since he went to sea when he was nine. Six years after that he was working in the mackerel fishing fleet and later entered the coasting business where he became a licensed master of sailing vessels. Later a captain's first class steamboat license was issued to him for the Atlantic coast. Before he joined the Life-Saving Service at the age of 36 he was a boat fisherman around Chatham. Not only had he years of experience with the sea and its moods, but he knew small boat handling, a prime requisite for rescue work. He was hired as surfman at Monomoy station and worked himself up to be number one in seven years.

They launched the smaller surfboat quickly and Ellis took the steering oar and after a hard pull down the inside of several miles they rounded the point and started looking for their captain, Marshall W. Eldredge of South Chatham, who looked like a man to be in charge, standing well over six feet, Eldredge weighed 220 pounds and was considered the toughest surfman who ever patrolled Monomoy Beach in the best and the worst of weather. He liked the feel of sand on his toes and so went barefoot until the calendar said December. In the early months of spring and the fall he refused to wear oil clothes, preferring to be wet rather than weighed down by them. He was appointed keeper of the station in August, 1899. Eldredge was married with three children.

He had seen the schooner-barge *Wadena*, loaded with coal and still aground on Shovelful Shoal with an American flag, union down, snapping in the strong southeast wind. It was still there, so he and his crew would row out into the rips and surging tide at the tip of Monomoy. When he saw his men and their oars rising and dipping he waded out where he could climb over the stern and take the steering oar from Ellis and head for the stranded barge, which they reached about noon.

It was a heavy sea and a dangerous place in which to handle a boat, so Eldredge and his crew rounded the barge to the lee side with the bow pointing toward the *Wadena's* stern. Close to the forerigging seemed to be the safest spot along side her 212 foot length.

The men on the barge-there were five of them-had had a frightening night from the southeast storm and its pounding waves. They had been watching the slow, careful approach of the lifesavers and their surfboat as it picked its way through the long swells and the dangerous rips. They were at the rail of the *Wadena* on the lee side when the rescuers arrived.

Captain Eldredge hailed the barge owner, William Mack, asking him:"You've got your colors in the rigging. What's the trouble?" Mack replied,"We had an awful night. The wind blew and those seas rolled in, and we pounded and thumped on the bottom and we thought we were going to pieces. We were afraid and still are, and we want to go ashore." "That's all right," said Eldridge, "but we're not coming aboard. It's too rough. Throw a rope over and lower yourselves down." The edge of the main deck was 13 feet above the water and one by one the men came down and were pulled into the surfboat until it was the barge captain's turn. Part way down he lost his grip on the rope and fell onto the thwart and broke it. Captain Olsen was a heavy man and he was put in the bottom of the boat along with two more. The other men were placed at the thwarts outside the oarsmen. The surfboat had shipped some water on the way out to the barge, and after this was bailed out, it was time to head for the beach. That is when it happened!

There was no one on the barge to cast off, so the bowman, Osborne Chase, cut the painter and at that moment the surfmen on the port oars were backing hard and the others on the starboard side were pulling. A big sea rolled in and shoved the surfboat ahead and out away from the protection of the lee of the barge. Then the tide rolling past the barge pulled her into the trough of the waves where the oarsmen were trying frantically to turn her toward the waves and into the wind. Before they could do that a heavy sea rolled over the surfboat. The bargemen, instead of remaining calm and as low as possible, panicked, jumped up and threw their arms around the surfmen who then couldn't row. Still in the trough the small craft was drifting and another sea broke over the helpless vessel, half filling her with water. Out in that violence of gale-force winds, tidal rips and long breakers on Shovelful Shoal the surfboat was buffeted several more times .

Keeper Eldredge and his surfmen were aware of the great danger and they instinctively knew they had to right the capsized surfboat in order to survive in the bitterly cold water. Twice they righted the surfboat and each time a heavier wave rolled over them, upsetting the craft. Again, thirteen men were clinging to whatever they could on the half submerged craft.

The five bargemen were washed away first, including the owner, William Mack, who at 29 was one of the youngest there. It was a bleak prospect for the eight lifesavers dressed in their heavy storm clothes and boots with enough weight to drag each of them down. Still they hung on until a heavy sea broke over the bow where Osborne Chase had his head barely above water. The wave pulled him away to his death. Chase was the first surfman to lose his grip as the overturned surfboat was wildly bobbing and tumbling in the violent water of the shoals.

With the thwarts submerged the only place to get a grip on the bottom of the hull were those long lapstrakes. But those were no guarantee of safety as another big wave slapped the surfboat and cascaded over Keeper Eldredge, washing him off. He frantically grabbed for the boat, missed and drifted away and sank out of sight. The next to go was Elijah Kendrick who became caught in the lines and mainsheets of the auxiliary sail that all surfboats carried to be used in a fair breeze in place of rowing. It was a rule that when not in use the sail was to be secured by lashing it to the thwart. That was not the case this day and the sail floated up right under Kendrick and his feet became entangled in the halyards and the sheets. It was a terrible burden for the former athlete; Kendrick had excelled in skating, swimming, running and baseball. With all his strength and agility he held on to the boat while furiously kicking to free himself, but that extra weight was too much and as his fingers slipped he turned and threw himself onto the mast, boom and sail in a bid to stay alive. Kendrick did not succeed; he floated a short distance and disappeared.

The storm pushed the overturned surfboat and the five men onto the shallowest area of Shovelful Shoal south of Monomoy Island. The surface of the sea here was worse than in the deeper water and the waves came booming in, breaking and pounding and foaming while stirring up the shallow bottom, making a violent mixture of sand and water. All the time there was the great noise of the sea.

The surfboat was almost completely submerged as it drifted into the most violent part of the tidal rip, carrying Nickerson, Rogers, Small, Ellis and Foye. Their situation was desperate. Five lifesavers barely hanging on to life. In that turbulent water three more men lost their grip and fell off to their deaths on the shoal, so Nickerson, Small and

Foye disappeared one by one in that ugly water. Of the eight lifesavers who had set out, because of a distress signal on a schooner-barge, only two-Arthur Rogers and Seth Linwood Ellis-were still alive. And just barely.

The sea was a little smoother on the other side of Shovelful Shoal and the half-submerged surfboat drifted that way, but the threat to their lives was not over. An exhausted Arthur Rogers was clinging to the bow. His heavy storm clothes were waterlogged and his exhaustion was so intense he could hardly raise his arms. The icy water and the pounding from the sea had taken his strength when a wave lifted him from the bow and carried him amidships where he feebly tried to hold on to the lapstrakes. His hands would not respond and the following wave lifted him off the wooden bottom and into the sea where he vanished. That left only Seth Linwood Ellis still alive.

The surfboat drifted slowly into calmer water and the centerboard slipped part way out, giving him a firm place to grip and with far less exertion. After a short rest Ellis pulled off his oil clothes, his jacket vest and his heavy storm trousers. Last he kicked off the heavy boots. He even thought about swimming back to Monomoy, which he believed he could do without those heavy clothes and boots.

At some distance from the *Wadena* and also aground on the shoals was the other barge, the *John C. Fitzpatrick*. On board was the barge captain, Andrew Welsh; a marine insurance underwriter who had come down from Boston named Captain Benjamin Mallows; and a Chatham maritime man, Captain Elmer F. Mayo, in charge of the wreckers who were busy lightering the cargo of coal. A fire had been built in the donkey boiler and when steam was up the pumps were started which also helped lighten the vessel. At one point Elmer Mayo was working on deck and he glanced over the port rail and saw a capsized boat with four men holding on to it. Then he recalled having seen a distress signal flying from the rigging of the *Wadena* earlier that day and after that the weather had closed in. When he saw the capsized boat he figured it must be a surfboat from the Monomoy Station which was drifting toward the *Fitzpatrick*. Captain Mayo grabbed a large wooden fender on the deck and heaved it over the side, hoping it would drift toward the men in the water. It did not. While still watching the drama in the water, Mayo saw three of the four men lose their grip and disappear. He had to do something!

On the barge was a fourteen-foot dory, hardly a vehicle for rescue in that sea, but Mayo saw it as the only hope available. There was still one man riding that overturned boat and Mayo wanted to get to him. The dory had earlier capsized and lost its oars and thole pins. It was in no condition to go anywhere. For thole pins, Mayo used a rasp, a serving stick and a couple of pieces of pine wood hastily found and inserted in the holes in the gunwale. Two old oars that were sawed off were found and tossed in the dory. Mayo ordered the dory lowered from the davits along side of the raised deck house aft. Captain Welsh said, "No, your dory won't live in that wild water, sir." Mayo ignored him and climbed in the dory and told his wreckers to lower it to the water which it hit without turning over. Mayo unhooked the tackles fore and aft, shipped the oars and pulled for the last spot where he had seen the overturned boat. Elmer Mayo had been around small boats for years and he knew how to work them and when he rowed out from the lee he took each wave in the bow and went over them one at a time and safely. There was another wave hissing and forming and ready to curl and he faced it with the dory's bow and she climbed up and over. He kept looking for Ellis on the bottom of that overturned surfboat. Then Mayo worked his way to the windward where a larger wave than usual was forming and he followed it, coming down on the crest and just keeping her steady and letting the ocean do the work. So that wave spent itself, but while on the crest Mayo saw Ellis again and not too far away. Another wave formed and Mayo held her steady and then he was on the crest and running down and there was Ellis about four or five feet away. When the lull began after the wave, Mayo pulled furiously on those stubby oars and backed up to the surfboat and there was Ellis who recognized him. Right along side and Ellis reached for the gunwale and pulled himself up and over and then into the bottom of the rescue dory. Now it was time to head for the beach and safety!

The wind was still strong; there was no break in that. Elmer Mayo continued to ride with the waves, some of them fifteen feet high that lifted the dory to the crest and then a rush down the other side toward the beach on the ocean side of Monomoy. The Atlantic was pushing that dory and its two occupants, and they weren't out of danger yet.

They were getting closer to the breakers near the shore and Elmer Mayo saw a man in the distance running toward them in the sand, but he wasn't nearby yet, so Mayo used his oars in the trough between waves to slow movement toward the beach, waiting for the right sea-the momentary fairly calm sea. And waiting for the man on the beach to be opposite them. The figure was getting closer now but the sea wasn't right. Soon up high on the crest of a comber Mayo saw surfman Francisco Bloomer from the Monomoy Station looking toward him and waving. Bloomer was opposite them. Now it was time to pull through the breakers to the beach!

In front the waves were still heading past them in an unbroken line like long rolling dark wet hills. In the trough Mayo lingered and with his oars held the dory back, waiting for the right sea. When the water is running in waves like that there are high waves-three,four, five- or more higher with a crest and a deeper trough followed in no particular sequence by a smaller one of relatively calmer water. The 40-year old Elmer Mayo with a world of experience in small boats patiently leaned on his oars, listening to the combers crashing ahead of him toward the beach, waiting for a moment-just a moment-of calmer water. Then it came. With a quick glance he saw it breaking under their stern and the small dory rose on the swell and there was Mayo furiously pulling on those short oars. They rode the crest into the beach where Bloomer had run into the shallows. Mayo jumped onto the sand and helped Ellis out of the dory. The wave rolled its limit up the beach, then receded, sucking at their ankles. Then the three of them with help from the next wave pulled the dory above the high water mark.

Because of the bravery of Elmer Mayo, Seth Ellis survived. He was the only one of those eight lifesavers who rowed out to the barge *Wadena* on a wild, raw day in March, 1902.

Elmer Mayo was always known after that as the hero of Monomoy following the saddest event in Cape Cod lifesaving: the death of twelve men in Chatham breakers. Twelve men down.

The only survivor of the Monomoy Tragedy, as it became known as, and his rescuer. Sitting is surfman Seth Ellis with Captain Elmer F. Mayo, a salvage man from Chatham. Both received government lifesaving medals. Ellis was promoted to keeper.

This is the altar in the Orleans Congregational Church for the funeral service for the seven lifesavers lost at Shovelful Shoal. At many Cape Cod churches there were similar services. The offerings were forwarded to the families of the seven.

"I never saw our boat handled better," Seth Ellis told people the day after the tragedy. "Captain Eldredge was cool and collected. All of our men were cool and not a bit of fear entered our minds. If Mr. Mack, the barge owner, who was the one least-accustomed to the sea, felt any fear, he had too much grit to say anything."

Ellis spoke of his friend, the keeper. "Captain Eldredge was true to his calling to the last. I shall never forget the look of agony on his face when he saw us all in the water."

The five Harwich surfmen who died left their wives as widows and a total of twelve children without a father. The other two, Eldredge and Foye, were from South Chatham and their loss left two widows and four children.

There was a great interest in the Life-Saving Service throughout America, as their stories had been told in national newspapers, magazines and through dispatches by the Associated Press and other wire services. Thus the Monomoy disaster was known all over the country. Expressions of sympathy were received through the mail and the telegraph to Harwich and Chatham. A number of memorial services were held in Cape Cod churches.

A drive for funds to help the families of the seven men raised more than $45,000.

The U.S. Life-Saving Service Report of June 30, 1902 stated: "By far the most distressing calamity to the Life-Saving Service during many years, and one unequaled by more than two or three in its history, was that that occurred on the 17th of March, 1902, near the eastern end of Shovelful Shoal, coast of Massachusetts, and resulted in the drowning of 12 persons, five from the stranded coal barge *Wadena* and seven from the crew of the Monomoy Life-Saving Station. The loss of the seven life-saving men who so nobly perished created everywhere a sense of profound sorrow. There was no more skillful or fearless crew on the whole coast, and since it appeared that the *Wadena* remained safe for days after the disaster, there was a general conviction that the men were practically a sacrifice-on the one hand to the needless apprehensions and senseless panic of the men from the barge, and on the other to their own high sense of duty, which would not permit them to turn their backs upon a signal of distress. `We must go,' said the keeper (Marshall Eldredge); `there is a distress flag in the rigging!' "

A dozen days later, on the 29th, the insurance underwriters floated the barge *John C. Fitzpatrick*. Tugs pulled her to Boston. The *Wadena,* still stranded on Shovelful Shoal, had her bottom torn out and she was a total loss.

Later in a ceremony at Chatham Town Hall, acting on behalf of the U.S. Treasury Department, Congressman William Greene presented gold Life Saving medals to Elmer F. Mayo and to Seth Linwood Ellis, who by then had been promoted to keeper at the Monomoy Life-Saving station. Ellis was 43 years old and had spent seven years in the Life-Saving Service-all of that time at the Monomoy Station.

The mother of William Henry Mack and his sister paid for a monument to be placed beside the Coast Guard station at Chatham. The two wanted the memorial to be erected near the beach at Monomoy Point opposite Shovelful Shoal, but the sands there shift constantly and that location was considered too unstable for anything permanent. Finally the U.S. Lighthouse Department gave its permission for placing it beside the twin beacons at Chatham.

The monument is an obelisk with an inscription on all four sides of the base. There are references to the fourteen men who were there: the two who survived and the twelve who perished.

On one side the inscription reads:

In Memory of

WILLIAM HENRY MACK
Of Cleveland, Ohio
Erected October 1903
By his loving mother and sister

Twilight and Evening Bell
And after that the dark:
And may there be no sadness
of farewell
When I embark:
For tho from out our Bourne
of time and place
The flood may bear me far
I hope to see my pilot face to face
When I have crost the bar.

The William Henry Mack permanent monument was dedicated in October, 1903 to all those who were at Shovelful Shoal that fateful day.

That stone memorial still attracts visitors as across the street, Shore Road Boulevard, the parked cars facing the ocean are lined up side by side every day.

Many of the guests look at the ocean and later come over to the other side and walk around the obelisk and read the four sides which are a permanent reminder of the sad ending so long ago on Shovelful Shoal of Twelve Men Down.

AFTER THE *WADENA* TRAGEDY

In April,1902 a coastal caravan of a tug and her barges were involved in an accident that sank a steamer in Pollock Rip Channel. It was 2:30 a.m. and the tug *Plymouth* was towing three schooner-barges loaded with coal for Portland, Maine when the last barge, the *Paxinos,* struck the steam collier *Williamsport* as the tow was rounding *Pollock Rip Lightship*. The length of those tows was a problem when turning as the tug is so far ahead of the barges.

The sunken *Williamsport* was discovered by the acting keeper, Seth Ellis, at 5:30 a.m. The surfmen launched and rowed to the *Williamsport*, arriving at 7 a.m. and found no one on board as the crew had gone to the tug *Plymouth*. The barge, *Paxinos*, was close by with a flag in the rigging union down and the crew in a small leaky boat. The lifesavers took the barge crew on board their surfboat and went to the barge and got up her sails and floated her to a place in the channel where they dropped anchor. The vessel, loaded with coal, was leaking very little. The tug *International* came and towed the *Paxinos* with her crew to Vineyard Haven. The *Plymouth* left for Portland, Maine with the other two barges of coal. The *Paxinos,* valued at $40,000, was only three years old, a product of Robert Palmer & Sons, of Noank, Connecticut, which built most of that huge coal fleet of the Reading Railroad Company of Philadelphia. Seth Ellis was the Acting Keeper and this was his first wreck report.

There were many situations of danger and bad seamanship that were made worse by alcohol. Sailors as a group had a reputation as heavy drinkers while ashore, but a number of ship's officers could not stay away from the sauce when they were at sea. Another example of that wrecked the three-masted schooner *Dora Mathews* sailing without cargo from Newburyport to New York.

One night she was beating down the coast past the Monomoy Station before a fresh east wind moving a rough sea. Her captain, Venson Bonsey, steered her toward the beach where she stranded. She was discovered at 1:30.

On the midnight watch in the tower, surfman Edwin Studley saw a schooner headed for the beach in a dangerous situation. Studley ran down to the water and immediately burned his Coston flare to warn the schooner off. When the schooner did not turn away, he burned another. The signals were seen by the crew on deck, but the captain would not take any notice of it. On the beach the surfmen shouted "Let go your anchors!" The crew said they heard it, but the captain being the worse for liquor would not let them do it. Shortly, the schooner came broadside to the beach and stranded fifty yards out.

Studley ran back to the station and notified Ellis who called out the crew, took the beach apparatus and headed for the wreck. It being low tide there was no need to fire a line, instead they put it on board with a heaving stick and took two men ashore. All the rest, except the drunken captain, walked ashore.

Ellis climbed on to the *Dora Mathews* and found the captain, as he put it, "very drunk." Ellis and his men lowered the sails and furled them and tried again to get the captain ashore. He said he would not come until he got ready. Ellis and his surfmen reached the wreck at 1:30 a.m. and stayed until seven that morning.

An hour earlier the Monomoy wreckers arrived and the captain had them removing everything that was movable-sails, anchors and other items from the deck. The value of which amounted to $500. The ship was twenty years old and had a high value of $15,000.

The next day Ellis took the crew to Chatham and got them free passes to Boston on the New Haven Railroad. That same day the owners came and one of them stayed four days at the life-saving station while looking after his property. The keeper's report is quoted in this accounting of the wreck, but does not include whether or not Ellis told the owners of the drunken condition of the captain as the cause of the destruction of their vessel.

Keeper Seth Ellis, in dark jacket, and his surfmen at the Monomoy Life-Saving Station in the late spring of 1902.

By late spring of 1902 it was finally decided at headquarters in Washington, D.C. to build and staff a life-saving station at Monomoy Point. Tuttle's urgings helped to create it. It was unfortunate he did not live to see it come about.

But come about it did and for the position of keeper the service chose Joseph C. Kelley, the Number One surfman under Captain Herbert E. Eldredge at the Chatham Station.

Kelley was twenty-nine years old and the youngest station keeper on Cape Cod, if not in the United States. Like so many others in the service, Kelley had great experience along the shores of the Cape. He had been a boatman and fisherman and a coastwise sailor.

Five years before his appointment as keeper at Monomoy Point, Kelley was assigned to the Peaked Hill Bars Station under Captain William W. Cook, a venerable lifesaver in that dangerous location. Kelley was there only a few months when he was transferred to the Chatham Station where the work was mostly rescuing seamen from schooners that were caught in northeast storms and stranded on Chatham Bars. Kelley very quickly was made Number One Surfman.

Joseph C. Kelley was the first keeper of the new Monomoy Point Station in 1902. Kelley, at age twenty-nine, was the youngest keeper on Cape Cod, if not in the United States.

He was skilled at handling boats in rough water-the prime requisite for being a lifesaver. Kelley's leadership abilities were the major reason for his getting appointed as keeper at Monomoy Point and the station there was manned on October 30, 1902. That was seven months and two weeks after the deaths of twelve men at the stranding of the schooner-barge, *Wadena,* on Shovelful Shoal off South Monomoy Island.

Within five weeks after the new station opened there were five wrecks. In each case the ships were saved by quick action by Kelley and his crew and all lives in peril were rescued.

There were three horses in the stable at the station. One of the horses, Susan, was a government animal. The other two were owned by surfmen. For many years keepers were allowed to rent horses, but were not provided them. Also Kelley's new crew had seven surfmen, rather than the usual six. Within the decade his crew numbered eleven, for reasons that we will explain. The station was located near the extreme south end of Monomoy Island, some nine miles from the twin lights at Chatham.

When the new station was approved to be built, it was to be a replacement for the older, established Monomoy Station to the north of the Monomoy Lighthouse. Part of that idea was to transfer the crew from there to the new station. After the Monomoy Disaster, with its death of seven lifesavers in March, 1902, the service decided to keep the Monomoy Station and proceed with the building of the one at the Point. Another reason could have been the distance between the Point and the Chatham Station which would have been one of the longest beach patrols on Cape Cod.

The new station featured large, comfortable rooms for the crew and those sheltered following wrecks. There were two boat rooms each with double doors and a wooden ramp in front. The station had three surfboats with one of these a self-bailer, the only one on the Cape. It was the habit of the service to furnish new stations with at least some of the latest equipment. There were two beach carts with the Lyle guns and breeches buoys and the rest of the apparatus. The usual lifecar was also placed in the boat room, although it was seldom used. The surfboats were the most important item, because Massachusetts was surfboat country.

The Monomoy Point Station was opened in October, 1902. Lifesavers had long thought a second station was needed closer to all those accidents near the Point. It was a long time coming. The *Wadena* disaster was the great impetus for its authorization.

The beach patrol north exchanged checks with the man coming south from Monomoy Station. Each of them walked one and one-half miles. The south patrol from the new station walked about the same distance, carrying a time clock which was activated by a key in the watch house at the extreme southern end of Monomoy Island.

A small schooner with a cargo of tar from Boston for New York ran into trouble off Chatham in a heavy sea on January 18, 1903. It was bitterly cold with a great deal of ice in the water and a strong wind out of the northwest. The two-masted *Emeline G. Sawyer* dropped an anchor some four miles north of the Monomoy Station.

Cape Cod Lifesaving Stations

Of the Outer Coast

At daybreak the morning watch discovered the vessel and thought she was in a bad place, so a watch was kept on her for fear she would get in trouble. After awhile she did.

The wind increased to a gale and the schooner could be seen rolling heavily. The anchor chain parted and her crew raised a part of her mainsail and jib and she started for the shore, finally striking on Common Flat Shoal and sinking. Ellis ordered the surfboat launched and the crew started a three hour pull to cover the four miles through the ice. What made it so hard was the continued icing up of the boat, making it difficult to keep the oars in the rowlocks. When the crew got within 500 yards of the wreck, the wind increased and the waves were higher and soon the surfboat was being driven astern. Ellis ordered the anchor dropped and they lay there trying to recover their energy enough to go on. Ellis said to his men,"We will have that crew or die trying."

Looking to windward, Ellis saw a surfboat coming from Chatham, and he thought it was from the Chatham Station. It turned out to be a volunteer crew from Chatham. Ellis urged them to go on and get the crew as their men were fresh. They did that and tried to row to Chatham but the gale stopped them, so both boats pulled for the Monomoy Station-an easier row. Once there fresh clothes, hot drinks and dinner helped everyone to recover from the cold. The crew of five stayed two days and then were taken to Chatham for a train ride home. The schooner was a total loss.

A number of routine rescues took place around Monomoy keeping both the Monomoy Station and the newer one at the Point busy. The summers saw a number of pleasure craft in trouble from the inexperience of their owners around the water. Many of these craft were sailed to Monomoy-an interesting place to cruise in good weather. Some of that time was the inactive season of June and July, so the keepers had to bring in one or more surfmen to help. With dories, catboats and other small craft frequently going adrift, the keeper had to advertise in local newspapers that they had been recovered. With the development of gasoline engines, many people were buying them, but their new technology was far from reliable. Many engines failed and still others ran out of fuel.

A schooner foundered and became a total wreck on Handkerchief Shoal the night of August 20, 1905 some four miles west of the Monomoy Point Station. The vessel, *M.C. Haskell,* was sailing for Rockport, Maine from New York with bituminous coal. The vessel ran into fog and the captain seeing two lights ahead thought they were marking the shoal. Although he was in doubt as to his exact position he thought he was in the channel. He kept on until about 10 p.m. when his schooner struck on the south end of Handkerchief Shoal and immediately began to fill and settle. Within three hours the water reached the cabin floor and the pumps could not keep up. The waves were breaking over her and suddenly she gave a great lurch and rolled on her side. The crew, with the exception of the mate, groped their way up the hatchways and took refuge in one of the boats. They cast off and made it fast to the jibboom. After dawn the life-saving crew from Monomoy Point discovered them, as the fog had lifted, and the surfmen reached the wreck and brought the five men ashore.

The mate was missing and the captain thought he drowned when the schooner sank. Kelley, the Keeper, obtained the services of a diver, who recovered the body from the cabin. The remains were sent to the home of the deceased for burial. When the *Haskell* struck at 10 o'clock the night before, there were men on patrol near Monomoy Point. If the captain had burned distress signals without delay, the surfmen could have reached his vessel before it rolled on its side, causing the death of the mate. There was time to rescue everyone. The crew of five were sheltered at the station and then taken to Chatham for the train home. Almost one year later the active season had just begun when the Monomoy Point crew went out to Pollock Rip Shoal where a coal schooner had grounded.

It was August 7, 1906 when the big *George V. Jordan* sailing from New York to Gardiner, Maine stranded in heavy fog. She soon filled with water and Kelley's men pulled for the stricken ship, rescuing her eight-man crew. It was fortunate she was seen, as those summer fogs off Monomoy can be thick. The *Jordan* was a total loss-$12,000 for the ship and $3,000 for the cargo.

The first big self-bailing, self-righting power lifeboat on Cape Cod was assigned to the Monomoy Point Station in 1907 and was kept in a house built expressly for it four miles north of the station at what was known as Inward Point on the ocean side of Monomoy Island.

This is the first self-bailing, self-righting power lifeboat undergoing tests in Provincetown Harbor. When the Service accepted the craft, it was the first on Cape Cod and was assigned to the Monomoy Point Station and placed in its own boathouse four miles north on the ocean.

The crew took turns living at the boathouse for periods of two weeks. A telephone call for the lifeboat day or night was answered at once. The doors opened and she was run out of the house on her carriage above rails that carried her to deep water. The crew of four started the engine and she sailed off to meet Captain Kelley and the others at a prearranged spot to carry them to the scene of a vessel in distress.

Attempts to keep the new power lifeboat nearer to the station failed owing to shifting sands that buried the tracks from boat house to water and prevented her launching.

Because of the need to keep four surfmen at the boathouse, Kelley had a total crew of eleven men, the largest group on Cape Cod.

A small, two-masted schooner with lumber from Calais, Maine for Hyannis and West Haven, Connecticut anchored during the night of January 4, 1907 in the channel near the southeast edge of Handkerchief Shoal. The weather was thick and a heavy rain was falling. The vessel, the *Alice T. Boardman*, last saw an aid to navigation, the *Shovelful Lightship*. After that all lights were lost in the heavy gloom, and the captain had to depend upon his judgment in running over and clearing the numerous small shoals in that area, a dangerous enterprise, because of the strong currents.

After tacking several times and failing to make the fog bell on *Handkerchief Lightship*, the captain decided to anchor until the tide turned and the weather cleared up.

A few minutes before midnight he let his starboard anchor go then furled his sails and set a watch. Around 4:30 a.m. in clearer weather it was discovered the vessel was dragging toward the breakers.

Instead of making sail and looking for a better spot, the captain ordered the port anchor dropped. It did not help. The holding ground was poor and soon the schooner was moving toward the breakers again. She soon struck the bottom and began to pound, moving up the shoal with the seas breaking over her, so the crew took refuge behind the masts then climbed into the rigging. Their situation was grave and they decided to launch the one small boat they had. They would row to the beach and safety.

As soon as the boat hit the water, one crew member jumped into the bow and then a wave knocked the boat under the stern and the crewman into the water. The tide swept the man away from the boat toward some floating planks and then into the breakers where he disappeared.

Finally someone entered the cabin and brought out an American flag which was placed in the rigging union down. A surfman in the tower at the Monomoy Point Station saw the distress signal and notified Kelley who rounded up his crew and they took off for the wreck.

When they arrived they came in on the schooner's lee side and took the crew of four off without a problem. Two days later the men were returned to the *Boardman* which was floated and towed to Hyannis assisted by a tug. The body of the drowned sailor was not found. About $500 of the cargo was lost from the deck.

Had the sailor who jumped into the boat stayed aboard his vessel, he would have survived as his crew mates did. But his fate may have convinced the others to avoid a similar adventure. Had they tried in that violent water they would not have had one chance in a hundred of success. One rule the lifesavers reminded shipwrecked sailors of over and over was to stay with your vessel and we will reach you with a seaworthy surfboat and get you ashore. A few sailors ignored this, coming to safety in their own boats and lived, but not many.

The last report by Keeper Seth Ellis at the Monomoy Station was dated December 18, 1907. Ellis retired after that and was replaced by Charles H. Howes who stayed on the job two years. In 1908 for the first time the service agreed to grant each surfman an additional $9.00 per month for rations. This was a big increase for the men as before they had to buy their own food from the monthly stipend of $65. They still had to furnish their own uniforms and foul-weather gear.

Most of Kelley's crew at the Point were decorating a Christmas tree on December 24th, 1908 when the lookout in the tower saw a big schooner in the breakers on Handkerchief Shoal. She was obviously in serious trouble and the keeper ordered the men at the boat house to launch the power lifeboat and come by the Point to pick up more surfmen. This they did with utmost speed. The swells were high and in that dirty weather and a bitterly cold westerly gale they could see waves cresting over the schooner *Harry Messer* and pouring white foam from her scuppers.

In those conditions that vessel would soon start to break up. The *Messer* was a stout ship and loaded with coal she had taken on at Baltimore for Boston, but with all that weight she was pounding heavily on the bottom from the seas. The lifesavers came in on the lee side and on the second try took off the crew of eight without a hitch and Kelley and his men got them back to the beach in the power lifeboat to the warmth of his station.

After dinner that Christmas Eve eight grateful sailors joined the surfmen at the Monomoy Point Station, singing carols around a decorated tree. The twenty-four cats who lived at the station were there, too.

Out in the water a mile and a half to the southwest the *Harry Messer* was dying with huge waves breaking over her decks. The storm was relentless. In a tangle of rigging wires and ropes the masts had fallen. Her deck houses washed overboard. The hull could not stand much more and it, too, finally broke apart in the gale, spreading her pieces over Handkerchief Shoal. By morning there was little left. The schooner was a total loss of $13,000 and $4,000 for the cargo of coal.

The men of the *Harry Messer* stayed at the Monomoy Point station through Christmas Day, and the following morning after a big breakfast left for Chatham where they took a train for their homes on December 26th. The tree with its decorations stayed up for three weeks.

During the summer of 1908 there were the usual emergencies with catboats, gasoline launches and other small craft. All the stations in the service had their share of these. However, there was seldom a risk of life during the good weather. It was frequently a case of retrieving boats adrift, or towing someone back to port. During the Thanksgiving season that fall a lumber schooner stranded on Handkerchief Shoal almost five miles west southwest of the Monomoy Point Station. It was just after midnight and the *Florence A.* was pounding heavily in a gale wind. The crew had taken refuge in the deck load of lumber when one of them burned a torch signal. It was seen by the lookout in the tower and the crew quickly launched a surfboat and pulled in the direction of the schooner.

The crew was taken back to the station where they were sheltered and fed for two days. When the wind calmed, the *Florence A.* was floated with the help of the lifesavers and towed to New Bedford.

The *Horatio Hall* sunk in Pollock Rip Slue after her collision with the *H. F. Dimock* in heavy fog, March 10, 1909. The passengers from the *Hall* went on board the *Dimock*.

After a patch was placed on her bow, the *H. F. Dimock* was pulled off the beach at Orleans and towed on March 14 to Boston for repairs.

From 1871 when the Life-Saving Service with Sumner I. Kimball was formed until the end of 1908 there were on all coasts and the Great Lakes a total of 18,411 maritime disasters. Of the 127,395 people on board the vessels, 1,004 were lost. Sheltered at the stations were 21,285 people. The value of the property at risk was $265,046,509, and the value saved $211,124,032. The dollar value of property saved was important to the service, as annually when looking for appropriations in Washington, the saving of property was an important talking point for additional money for continued expansions.

The end of 1908 marked the 37th year of the United States Life-Saving Service and all of it under one general superintendent: Sumner Increase Kimball from the State of Maine.

The Orleans surfmen with their keeper, James H. Charles, inside their boatroom after their rescue of passengers from the *Hall*. Surfman Edwin P. Ellis, on the left, was given a silver pitcher by a grateful passenger from the *Hall*. Keeper Charles is seated next to Ellis.

A major marine disaster off Monomoy involved a collision between the Boston-New York steamer *H. F. Dimock* of the Metropolitan Line and the New York to Portland passenger steamer *Horatio Hall*, of the Maine Steamship Company. In the fog the *Dimock* sliced into the port side of the *Hall* about two-thirds of the way from her bow. The collision occurred in Pollock Rip Slue. The passengers on the *Hall* were transferred to the *Dimock*. The *Hall* sank with her deck houses above water and a huge gash in her side. The *Dimock* was in danger of sinking herself with damage to her bow section and she sailed north and was beached south of the Orleans Station. Her crew remained on board.

Surfmen from the beach patrols, north of the station and south of it, heard the signals of distress-short blasts on the steam whistle over and over-and Mayo on the south saw the *Dimock* heading for the beach. He ran a half mile to warn the station. The big surfboat was housed a quarter mile away and Captain Charles and his men ran to it and launched into the rough sea. Once at the *Dimock* the lifesavers started taking off people from the *Hall,* including five women. All told thirteen trips were made to rescue sixty-seven of the passengers and crew from the *Horatio Hall*. The *Dimock* had a patch put over her damage on the starboard side of her bow and a number of tugs pulled her off Nauset Beach at high water four days later on March 14, 1909.

The power surfboat from the Monomoy Point Station was used to take underwriters to the *Hall* several times. Wreckers from the Point stripped the *Hall* of most things movable.

Lifesavers from Nauset, Old Harbor, Chatham, Monomoy and Monomoy Point worked the disaster for many days. It was a cooperative effort of all those stations. The *Hall* was right in the narrow part of the channel. Ships passed closely which was the main reason for that collision and for others through the years. The Government removed her with explosives as her position was a menace to the heavy traffic through there.

Two days after the collision at the Slue with the sinking of the *Horatio Hall*, a schooner was sinking four and one-half miles southeast of the Monomoy Point Station. The vessel, the *Fred A Small*, became stranded and started flying her ensign, union down, in the rigging. The lookout saw it and Kelley called out a crew who reached the vessel and went on board.

The surfmen helped with the pumps, remaining on board all night. Kelley and his men returned to the station and telephoned for the Revenue Cutter, *Gresham*, and a tug. The lifesavers returned to the schooner and stood by while the other rescue vessels tried to pull the *Small* off the shoal. They could not and the crew was taken to the station and later left for their homes. The schooner broke up in the next storm, a total loss. During this time some of Kelley's crew were involved with the wreck of the *Hall.* The cutter *Gresham* took part in many rescues often because she was asked by lifesavers, sometimes because while on routine coastal patrol she came across a vessel in distress.

It happened on a bitterly cold day in winter with the wind blowing a gale from the northwest, the sea running high, and the temperature below zero. The lookout at the Monomoy Station saw a schooner several miles off shore flying an ensign in her rigging, union down.

The lifesavers launched their surfboat and pulled for the vessel, reaching it in an hour. It was the *Rhoda Holmes* with coal for Rockland, Maine. She had set out from Vineyard Haven the day before with a favoring wind. After rounding Monomoy Point she encountered high winds from the west which damaged her mainsail and foresail. She ran back to the vicinity of *Pollock Rip Slue Lightship* where she set the distress signal in her rigging.

The lifesavers, under their new keeper, F. E. Hamilton, reached the *Holmes* at 9 o'clock in the morning of February 10, 1912. Her captain was fearful, because of the weather, so the surfmen stayed on board. The cutter *Gresham* arrived at 1:40 that afternoon. She had already taken in tow another ship in trouble, but she passed a line to the *Holmes* and began to tow both vessels to Boston.

After the towing started the crew of the schooner, being in need of sleep, turned in and left their vessel in charge of the lifesavers.

It was a slow journey and about 4:30 next morning between Highland Light and the buoy at Peaked Hill Bars, the hawser running to the *Rhoda Holmes*, the last vessel in the tow, parted.

Signals were made to the *Gresham*, but the sea was so rough that she could not pick up the schooner without cutting loose from the other vessel. So the *Holmes* with the lifesavers and her own crew on board was left to her own resources. With much difficulty the lifesaving crew patched and hoisted the foresail and ran her off before the wind. She had fallen into the trough of the waves when the cable parted and was rolling badly. In the violent water her rolling caused a cabin lamp to fall, setting a fire. The blaze was fought successfully and quickly.

The lifesavers were sailing the ship and they ran her before the gale southward to between Cahoons Hollow and Pamet River Life-Saving Stations.

By this time snow was falling and the *Holmes* was icing up badly on the deck and in the rigging from spray hitting the vessel. Everyone who could be spared from working the vessel was busy breaking ice. It was reported that as much as twenty tons of ice were loosened and gotten rid of.

The *Gresham* arrived before 11 o'clock that morning and took the schooner in tow for Provincetown.

The captain of the *Rhoda Holmes* said the lifesaving crew preserved his ship from destruction, but for their presence on board she would either have been burned or driven onto the shore of Cape Cod.

An old two-masted schooner from Rockport, Maine lost her anchors and stranded on Monomoy Beach north of the Monomoy Station when her crew abandoned her on September 2, 1912. The surfman on the north patrol discovered her at 1:20 in the morning and burned a Coston signal and got no reply. The *Silver Spray* was high up on the beach when found. Surfman Nickerson saw the schooner and he made for the halfway house and telephoned the station. Several others arrived and searched the vessel. They found her boat gone as well as all clothing from the cabin and her compass. They hauled down her mainsail and left, leaving one man to guard the property until an underwriters agent arrived whom they had notified.

Later that morning the keeper was informed that the crew had been picked up by a passing vessel, the *S.S. Dorchester*, west of the *Handkerchief Lightship*.

The wreckers stripped her of rigging and whatever else they could carry. She was forty years old and there was little worth saving. The hull was strained and worth nothing.

The use of the beach apparatus and breeches buoy was rare on Monomoy Island, but it was used to bring a crew ashore from a schooner in January, 1914. The vessel, the *General Adelbert Ames*, stranded on the southeast part of Monomoy Beach at 2 a.m. on the 21st.

A surfman on beach patrol saw the schooner and men with flaming torches on the deck. He telephoned the Monomoy Station from the watch house at the point.

The keeper at the Monomoy Point Station called, asking in what direction the vessel was. The Point Station sent a surfboat, but the sea was too rough and they couldn't get near enough to the wreck.

They called again and asked that the Monomoy crew come down and help with the beach apparatus. Shortly, both crews met on the beach opposite the wreck, which was in a bad position with high seas breaking over her near the beach and her cargo of laths washing overboard.

The beach was covered with water at high tide and there was one small place where they could bury the sand anchor. It was a difficult situation for using the breeches buoy. In fact, four attempts had to be made firing the lead with a line before one worked. One shot parted; one fell short after snarling and one other went clear of the beach before the last shot fell across the schooner and the rescue was finally made ready.

There were eight men on the *General Ames* and they were removed one at a time with the breeches buoy working perfectly.

Captain Kelley took the rescued sailors to his Monomoy Point Station for food and shelter. The Monomoy crew stayed and helped load the beach cart after which they returned to their place.

The *General Ames* went to pieces in the high surf on Bearse's Shoal and was a total loss. Some of the laths were recovered as was frequently the case when lumber schooners stranded.

The Federal Government brought the United States Life-Saving Service and United States Revenue-Cutter Service into a new organization, The United States Coast Guard, in an act of 20 January 1915. By this act the lifesavers finally won what they had never had, a retirement system.

They were civilians as lifesavers, but now they would have military rank. The superintendents of the various districts would become commissioned officers, the keepers would be appointed warrant officers and surfmen would be enlisted men with the Number One Surfman a petty officer-all ranks in the new Coast Guard. The General Superintendent, Sumner Increase Kimball, had been in that position for the entire forty-four years of the Life-Saving Service's existence.

On May 29, 1915 the last wreck report by the Monomoy Station keeper for the United States Life-Saving Service before it became the United States Coast Guard was of The Maine lumber schooner, *Mary E. Pennell* of Machias.

At 5:45 in the morning she stranded on Pollock Rip Shoals and was leaking badly. Within a half hour she was flying a Union Jack upside down.

Roger Cahoon, the acting keeper, brought his men in a surfboat to the vessel where they helped man the pumps. A signal flag was hoisted for the revenue cutter to assist. When she arrived a hawser was run from the cutter to the schooner. It was high water and the vessel came off.

The *Pennell's* captain wanted the surfmen to stay with his vessel for the haul to Vineyard Haven, so they did. The surfboat was tied to the stern of the schooner. When the tow reached repair yards at Vineyard Haven Harbor, the surfmen and Cahoon stayed overnight. The next morning the cutter towed the surfboat back to the beach on Monomoy near the station.

The story of the schooner *Pennell* on Pollock Rip Shoals was the last official report from the Monomoy Station to the United States Life-Saving Service. From then on in 1915, it was the Coast Guard.

THE LIFESAVERS OF CUTTYHUNK

The men of Cuttyhunk went to sea as watermen and they traveled far and wide not only as coasters but as deepwater men, too. In the early days lobstering was the summer occupation. Whereas in the spring and fall men fished for mackerel and also in the fall fishermen went after cod usually southwest of the island in the western end of Vineyard Sound. In the colder weather months the lobster boats went dragging with rakes for quahogs.

The sea supported them. It was their source of income. It was their livelihood, their life. When the whaling industry was the principal business of New Bedford with the occasional clipper ship from or to the West Indies or China, Cuttyhunk-opposite New Bedford-was the perfect spot for sighting incoming ships in need of a pilot. Some local watermen became pilots while others moved there because of a strong need for their services. These men could often be seen with telescopes, called spyglasses, and with binoculars searching the horizon for ships in need of piloting. They would climb Lookout Hill, as it was called, and search the water for business sailing through Buzzards Bay or Vineyard Sound.

Vessels sailing along the coast were counted in many places. In the year ending June, 1830 the number of vessels passing through Vineyard Sound was 12,603. Twenty-one years later in 1851 in a three-month period 6,676 vessels passed Cuttyhunk. If that flow of tonnage continued the annual total of vessels would be 26,704 or more than 70 ships each day.

Cuttyhunk has always been next to a water highway. Its watermen, pilots and fishermen, were involved with rescuing sailors whenever ships were in danger in that area. The Massachusetts Humane Society built a lifeboat station there in 1847, and later added another at the opposite end of the island. The United States Life-Saving Service built a station on the peninsula near Canipitset in 1889 during one of its periods of expansion along the Massachusetts coast.

The first attempt to aid mariners was the building of a lighthouse on the western end of Cuttyhunk in 1823. The keeper and his family lived in the government residence adjacent to the light.

The winter gales drove many a vessel ashore on Cuttyhunk and onto Sow and Pigs reef to their destruction. Another area to the east was a place of great danger to vessels called "The Graveyard". That was the shoreline along Vineyard Sound of Cuttyhunk, Nashawena and Pasque. Before the federal government's Life-Saving Service, the local men, all volunteers, made many a daring rescue along those rocky shores in usually horrible weather conditions. The Humane Society was a more organized effort with a boathouse, a leader and attempts at regular training sessions.

The earliest shipwreck recorded was a schooner driven ashore in the War of 1812 by British men-of-war vessels.

The reluctant entry of the United States Government into the field of life-saving changed that entire area greatly and for the better.

The Cuttyhunk Station received an unusual request February 24, 1889 from the captain of the schooner *George and Albert* of New York. The captain had employed local wreckers to load logwood from a vessel that had been wrecked on Cuttyhunk previously. The captain wanted to finish loading so as to reach New Bedford before dark. Extra hands were needed, so the keeper of the station turned out with his surfmen to help. It was not in their usual line of work. It was, however, helping the maritime community.

There were many catboats owned on Cuttyhunk and the other Elizabeth Islands. Their owners sailed them for most of the year even in times when they should have stayed in harbor. In December, 1891 two cats were in trouble. On the 18th one was overloaded and sank. It took five days for the surfmen to locate and raise her, returning the boat finally to her owner. The last day of the month, New Year's Eve, another cat stranded while attempting to enter harbor. She was seen and reported by the lookout in the tower. The crew carried out her anchor, warped her off and piloted her to a safe anchorage.

Less than three months later in March, 1892 the British schooner, *Rob and Harry,* became dismasted three miles west of *Hen and Chickens Lightship* in a west wind that turned to gale force and drove the vessel toward the east and Cuttyhunk where she repeatedly struck rocks and finally rose and fell in the breakers about three-hundred feet from the western shore.

On the Western end of Cuttyhunk the first lighthouse was built in 1823 and was replaced twice. This is the third and final light.

It was at least two hours before the stricken and half-sunk schooner was sighted by A.G. Eisener, the light keeper at Cuttyhunk who notified the lifesavers at the station. Eisener, in addition to his duties as light keeper, was in charge of the Humane Society station Number 43 at the western end of Cuttyhunk. He sent messages to society members, all volunteers, about the wreck and also notified keeper David Bosworth of the U.S. Life-Saving Station on the east end of Cuttyhunk.

The Humane Society's large surfboat was brought overland to the beach opposite the wreck and launched with Eisener in charge of his volunteer crew. They reached the wreck through high seas where they rescued the master and the mate. Before the schooner's other two crewmen could enter the surfboat, a huge wave crashed over the rescue craft and tore away some of its oars while rolling it half way to the beach. The surfboat was half filled with water and had several holes staved in her from the beach rocks.

In the meantime David Bosworth arrived on the scene two miles west of his station with the beach apparatus which had been pulled there by two mules. Seeing that the wreck was too far off shore to reach with the Lyle Gun, Bosworth launched a small surfboat into the towering surf and twice the seas threw it back, the second time creating a huge hole in her bottom. The situation on the wrecked schooner was desperate, so the rescuers turned to patching the society's large boat that had been staved and covered the holes with canvas. When the lifesavers rowed back to the wreck they found only one of the two men alive.

The dead man was the steward, Ephraim McNeil, who had frozen to death. McNeil's body was gently lifted over the side and into the surfboat. The other crew member and some of the lifesavers came over the rail and joined the others in the surfboat which was rising and falling alongside in the high seas. They slowly picked their way through and around the numerous reefs off that shore and landed safely.

One of the many wrecks around Cuttyhunk stands out as the grimmest for the islanders, a personal disaster for local families. During a violent gale February 24, 1893 the brig *Aquatic* stranded on Sow and Pigs reef in the early evening. At about 8:20 she was seen by Willie Eisener, the son of the lighthouse keeper, A.E., referred to all over Cuttyhunk as Captain Eisener. Willie shouted to his father, "ship ashore!" His father sent a message to the Life- Saving station at the eastern end of the island.

Shipwreck news spread fast and soon many islanders were along the shore peering through the gloom of the bitterly cold night to try to see the endangered vessel. Shipwrecks usually brought out everybody, even on horrible nights like this one.

Josiah Tilton was reading and trying to concentrate on the story while the wind roared around his house. To him it was a wild night; at about twenty minutes of nine he heard a violent knock on his door and Willie Black entered, breathless from running and blurted out that there was a sailing ship aground on Sow and Pigs.

Black went on to spread the word and Tilton left by himself for the Humane Society boat house on the west end of the island. He was alone when he reached there, but one by one the other volunteers arrived. Captain Tim Akin was the last volunteer to get there and when he finally did all the others were ready to start for the wreck.

On Cuttyhunk and elsewhere there was a strong rivalry between the surfmen of the U.S. Life-Saving Service and the volunteers of the Humane Society. It involved who was first to reach a wreck and save the most survivors while demonstrating the greatest skill. There was a lot of pride involved.

The surfmen from the life-saving station had not arrived when the Humane Society had gathered. The lighthouse keeper, Captain Eisener, told Tim Akin and his volunteers to wait for the life-saving crew from the east end who were dragging their surfboat toward the scene. Tim Akin did not want to wait and he chose five men to go with him: his brother Fred Akin, Josiah Tilton, Isaiah Tilton, Hiram Jackson and Eugene Brightman.

Despite the plea from Eisener that two surfboats would be better than one, Tim Akin shouted, "Man the boat!" They shoved off and after a hard pull were opposite the stranded ship in the breakers. In the long swells Captain Tim often turned to the others and said, "This is all right, this is all right."

Josiah Tilton was the only man who survived. He later told how he turned to Fred Akin and asked, "I don't think we will get there, do you?" Fred looked toward the huge breakers and replied "I don't know."

When the surfboat was a few hundred feet from the brig, Captain Akin shouted to get the anchor ready so we would have something to pull her back by. A much larger wave crested along side the boat and it flipped the small craft over, throwing everyone into the sea. Tilton who couldn't swim a stroke was carried by a wave to the side of the brig. He was bobbing in the water near the stern and a sailor ran aft and threw him a line which he passed under his arms and he was lifted up to the deck. Tilton saw Tim Akin crashed against the hull with a terrific impact that stunned him and he just washed away and disappeared. The others had vanished before in that wild water.

Tilton was led into the cabin and given dry clothes. Within an hour the pounding of the sea had opened the hull and she was filling with water, forcing everyone onto the deck. When the sea started breaking over the deck, Tilton and the ten crewmen from the *Aquatic* climbed into the rigging for safety. Tilton wrapped himself in the foretopsail for protection from the gale and bitter cold. He and the others spent about twelve hours hanging on to rigging in the brutal weather when they saw the tug *Elsie* from Vineyard Haven pulling the lighter *Aid.* Then the life-saving crew launched and pulled out to the lighter.

When the surfmen from the station reached the lighter they started working on the deck with the beach apparatus. The tug dropped the lighter astern and allowed her to drift near the breakers where she was stopped and a line fired across the wrecked vessel and a breeches buoy rigged. Tilton and two crewmen rode the buoy to safety one at a time, striking the water fifteen feet from the lighter when the line sagged. Each of them was up to his neck in the icy water when he was picked up by the surfmen who were standing by in the surfboat.

The lifesavers then rowed a big dory from the lighter through the breakers and under the bow of the *Aquatic* where the last crewmen-seven of them-came down the spring stay to the martingale and then jumped into the dory. The lifesavers rowed them out of the breakers to safety.

The U.S. Life-Saving Station crew made a valiant effort to reach the *Aquatic* by launching a surfboat and rowing out around Cuttyhunk on the Buzzards Bay side. The men were coated with frozen spray after a difficult pull of four hours to reach the breakers around Sow and Pigs where they stopped short of the huge waves after seeing how dangerous it was. Keeper David Bosworth ordered his crew to row for the shore at Wash Pond where a crowd of onlookers helped haul the surfboat out of the water.

Bosworth sent for his big lifeboat, which arrived at dawn. The sea was just too violent in the heavy gale, so Bosworth waited and later hailed the tug *Elsie* and the lighter when they steamed past. The Lyle Gun and breeches buoy were taken aboard the lighter and the rescue of the thirteen men followed.

The death of the five islanders left twelve children fatherless. Tim Akin's children were the worse off as their mother had died two years before the *Aquatic* disaster. Hiram Jackson had three sons and Isaiah Tilton had three children. Fred Akin and Eugene Brightman were not married.

News of the loss was spread through the wire services and America responded, collecting $30,000, with another $1,000 from the Canadian government. The money was not administered properly and one way or another it disappeared quickly. The five youngest children of Timothy Akin, Jr., with no one to care for them were placed in institutions. When the oldest daughter, Henrietta, married later on she gave a home to her younger brothers and sisters.

Over a few weeks the five bodies washed ashore on Cuttyhunk and were buried.

The tragedy of the wreck of the *Aquatic* happened because of pride or a sense of rivalry, but the trip that Tim Atkins insisted on making was really unnecessary, because the shipwreck offered refuge for its twelve-man crew, all of whom survived.

On October 11th of the following year, 1893, the American schooner, *Ada A. Kennedy*, was wrecked on Sow and Pigs Reef. She was seen by the beach patrol. Before the lifesavers could row out to the wreck, the crew landed in the schooner's boats after which the surfmen took them to Cuttyhunk Harbor and got them free transportation to New Bedford on a local steamer that ran regular ferry service to the island. The ship's master stayed on the island after his crew left, then he, the master, was taken to New Bedford on the revenue cutter *Dexter* which the keeper signaled as she was passing Cuttyhunk. The surfmen brought the schooner's boats into the harbor on the east end and hauled them up on the beach.

The *Ada A. Kennedy* went to pieces on the reef in the next storm. Nothing could be salvaged.

It was only five days later when another American coastal schooner misstayed close to shore and almost came to grief. Almost The vessel, the *Pascamansett,* had to quickly drop anchor to avoid stranding in the shallows. Her captain was temporarily disabled when he injured his ankle, leaving only one other person on board to work the ship.

The surfman in the tower on lookout spotted the schooner and gave the alarm. The lifesavers rowed out in the surfboat and slipped the anchor, or cable, and worked the schooner off shore, after which they recovered the anchor and put it on board. After the surfmen reefed the mainsail, the vessel proceeded to her destination.

Around Cuttyhunk there was at least one boat that ran a ferry service for passengers, mail and freight from New Bedford to the islands.

The *Helen Augusta* was the first followed by the tug *Charlie* in 1877. The steamship operation was taken over after 1885 by the New Bedford Towboat Company. They put the tug *Cygnet* of their fleet on the service where she ran until 1906. *Cygnet* was 62-feet long and could carry one-hundred and twenty-five passengers. She had to anchor outside Cuttyhunk Harbor as the channel into the harbor had not been dredged. Outside the harbor there were catboats that carried passengers to and from *Cygnet* along with the mail and freight.

The lifeboat station is the building on the left with a larger lifeboat on a cradle with rails to use in launching. The larger lifeboats always had to be kept close to the water.

Captain Darius Weekes and his surfmen before their beach apparatus and the boatroom at the station in 1895. The Life-Saving Service had built the station six years before in 1889 when it placed David Bosworth in charge. Weekes took over later.

During the inactive season of the summer of 1902, a schooner from New York to Salem, Massachusetts with coal stranded in dense fog on Sow and Pigs reef. It was early evening of July 2nd and the keeper of the life-saving station, went to the village to round up some volunteers to man a surfboat and a dory of the Humane Society.

When he had a sufficient crew of volunteers they rowed out to the schooner *Joseph Oakes* of Deer Isle, Maine, arriving at nine o'clock that night. The vessel did not seem to be too badly off; her captain wanted the keeper to help refloat her. The wrecking gear to do that was not available. The captain of the *Oakes* and his four crew members were brought ashore where they spent the night at the lighthouse.

The next day a strong wind came up and the schooner was driven farther onto the reef where she sank. Arrangement was made with a local wrecker to strip her of anything movable, and it was done.

A year later another vessel, the *William L. Elkins*, stranded on Sow and Pigs Reef and her captain ordered his crew to abandon ship which they did in their own boat, rowing to the life-saving station where they were given food and shelter. The following day, July 31st, they made arrangements with local wreckers to strip the schooner. Later that day the five crewmen took the afternoon steamer to New Bedford.

A major rescue effort was undertaken during dirty winter weather in February, 1899 when the American steamer *Admiral Dewey* went aground near the Cuttyhunk station at three a.m. The beach patrol spotted the vessel near the west end of his beat and fired a Coston flare. The surfmen hauled their boat along the shore on the ice and found a clear spot to launch and then rowed to the *Dewey*.

The stranded steamer was on her way to Boston from Jamaica, carrying eighteen passengers, fifty crew and a cargo of tropical fruit. The surfmen stayed on board until dawn, then went ashore and sent a message to the Gay Head Station, asking that a tug be sent from Vineyard Haven. Around noontime the tug, *John J. Hill*, arrived from the Vineyard and the lifesavers helped the passengers get aboard with their baggage.

The crew of the *Dewey* decided to stay with their vessel and so the lifesavers went back to the station. Around five o'clock the surfmen were hailed from the *Dewey* and told that some of the crew wanted to go ashore. The surfmen returned and took two boat loads of crewmen and two loads were brought ashore in the steamer's boat, making thirty-five all told who were landed at the station.

The following day, February 16th, the Revenue Cutter *Dexter* appeared from the east and her captain offered assistance. The surfmen transferred twenty-three men from the life-saving station to the *Dexter* and she carried them to New Bedford. The other dozen men were rowed from the station to the steamer *Admiral Dewey*.

Then the surfmen ran a line from the shore to the stranded vessel in order to have communication with her in case the weather turned sour and make it possible to rig a breeches buoy for rescues.

Later tugs and lighters arrived and within two days the tugs hauled her off, although her hull was badly damaged. The crew who stayed at the station overnight were fed and provided shelter.

Seven months later the lifesavers at Cuttyhunk responded to a call for assistance, because of a domestic animal. In this unusual case the schooner, *Mary and Belle*, was beached by her skipper in order to take a horse aboard. The thinking was to wait for high water and then float off. It did not work that way. The skipper raised a signal and the surfmen responded by laying out anchors and then working the schooner off, with the horse on deck.

A catboat was lying at anchor on October 20, 1899 about three miles WNW of the station. When the wind freshened, the boat made no effort to head for harbor. The keeper decided something was wrong so he called out the surfboat and on the way out her mast was carried away. When the surfmen reached the craft, finding no one on her, they hove up the anchor and sailed into a safe anchorage and returned to the station.

Some hours later the owner of the catboat showed up and told them he was about to get under way when the boom knocked his hat off. He launched his skiff and took after the hat, breaking one of his oars and then being unable to return to the catboat because of the high wind and tide. The man finally managed to land on the west end of Cuttyhunk and walked back to the life-saving station. The storm raged for two days and the sailor stayed put at the station where he was sheltered and fed. When the wind abated he sailed away in his catboat.

This happened in October when there are some fairly pleasant days to go for a sail. This sailor was fortunate. Others took greater risks by going out alone in catboats in the winter. Lifesavers risked their lives on many occasions rescuing such adventurers.

Early in March of 1900 the American schooner *Harry L. Whiton* stranded near Dumpling Lighthouse north by east of the station. The surfmen boarded her and first laid out an anchor. When the vessel started taking water they assisted at the pumps. A little while later a tug showed up from New Bedford and the schooner's master made a contract to float his vessel. The surfmen assisted by heaving on the windlass, and after the schooner worked free, they stayed with her, pumping her out all the way to a New Bedford shipyard.

One week later the sloop *Helen* stranded about two-hundred yards from the station. The surfmen rowed to her and laid out an anchor and hauled her afloat at high water, then worked her to her moorings in Cuttyhunk Pond.

A few days before the inactive season began that year, 1900 an American schooner, the *Annie E. Rickerson*, stranded after midnight on the infamous Sow and Pigs Reef, about three miles west of the station. The surfmen reached her and helped pump, but were unable to lay out an anchor because of the heavy sea running. Wreckers from Cuttyhunk made an agreement to float the vessel, but were forced to abandon the job. Toward night of the next day it was decided to take off all hands with their clothing and effects and land them at the station. This was done by the surfmen who were assisted by a small steamer owned by the wreckers. Not long after the men reached the station, the schooner was driven off the reef by a huge wave and began drifting to leeward. A passing steamer saw the drifting schooner and picked her up and towed her into New Bedford Harbor for repairs. The crew of six from the *Rickerson* stayed at the station one day and then took the steamer for New Bedford.

Less than a month later in June, during the inactive season when no surfmen were on duty, the American schooner, *Isaac H. Tillyer*, stranded on Cuttyhunk some three-hundred yards SSE of the station. The keeper assisted the crew to land their personal effects and the vessel's furniture. Her master made an agreement with local wreckers to float her, but bad weather set in and she was abandoned and became a total loss. The master and several of his crew stayed at the station several days.

During an especially thick fog in the winter of 1902, surfboat teams from Cuttyhunk and Gay Head rowed to the scene of a stranded steamer on the southwest part of Cuttyhunk Island. It was early evening on March 29th when the American vessel *Indian* grounded, breaking her shaft and losing her propeller. When that happened she started to fill and that put out her fires, and, having no steam, she was unable to sound her whistle. She could not make a visible signal of distress because of the thick fog.

The following morning the fog had cleared and she was sighted by the patrol who called up the station on the watch house telephone. At 5:30 the surfboat launched at Cuttyhunk Station and they pulled for the wreck scene. In the rough sea it was a hard pull, but they finally reached the stranded *Indian* but were unable to board her as tremendous combers were breaking constantly alongside her.

Finally the keeper backed his surfboat as near as good judgment would permit. The steamer's master threw the keeper a can of dispatches and asked that they be sent immediately. The surfboat went to shore where the keeper sent telegrams for the master, returning to the steamer by 10:30.

When the Cuttyhunk boat arrived back at the wreck the Gay Head boat was already there, trying to get close for a rescue. The seas were too rough and the Gay Head boat withdrew to wait for calmer water.

The Cuttyhunk boat decided to stand by for a favorable chance to move alongside. Around three o'clock that afternoon the boat made a dash for the ship and took off the three passengers: one woman and two men. The surfmen pulled clear just in time to escape a line of breakers that struck with terrific force. The master and his crew decided to stay on board for the night, so the lifesavers pulled for the shore and landed the three passengers within a short distance of a hotel. It was five o'clock.

The following morning at daybreak the crew returned to the wreck where they were joined by the Gay Head crew and the steam tug *Mercury*. The wind had died down and there was a moderate sea and the two surfboats transferred the twenty-six man crew to the tug. The mate was taken back to Cuttyhunk where he could keep a lookout for the wreck. The baggage of the crew was transferred to the tug, also.

Wreckers had been hired and on many days the lifesavers returned to the steamer to help in any way they could. The bottom of the hull was found to be seriously damaged, so the pumps would do no good.

Most of the cargo was lightered over many weeks, and the *Indian* was finally floated on May 15th, forty-seven days after she stranded at Cuttyhunk Island.

A yachting trip before Christmas in December 1903 could have been dangerous for a party of six caught in a strong wind and a choppy sea. It was three days before Christmas and the sloop *Helen* was at anchor on a lee shore and in danger of stranding. In response to a signal for help, the crew launched and pulled to the sloop a half mile northwest of the station. They arrived and hove up the anchors, made sail, and worked her to a safe anchorage and landed the six people on board, a very grateful group.

A month later in January a tug, towing a barge, stranded in a thick snowstorm on Sow and Pigs Reef. The barge was anchored and the crew of the tug went on board her.

The next morning, January 29th, the vessels were spotted by the patrol who telephoned the station from the watch house. The life-saving crew pulled the surfboat across the island through heavy snowdrifts and launched. They pulled to the barge where they picked up the captain of the tug, the *G.M. Winslow*, and took him ashore where he could telephone for assistance.

Soon a tug appeared and the lifesavers took the master out to his vessel where he surveyed his stranded ship. The lifesavers gathered the personal effects of the crew and returned them to the rescue tug which then took the barge in tow.

The stranded tug, the *Winslow*, proved a total loss.

With their U. S. Life-Saving Station in the background, the Cuttyhunk surfmen pose for their photograph in 1903. The keeper, Darius Weekes is on the left. The training wreck pole is at the right rear.

The Cuttyhunk Station from another angle. The flags and the surfboat indicate this is a Sunday visiting day in the summer.

Catboats seemed to get in all kinds of trouble. The *Themis*, was anchored near a ledge of rocks August 31st, 1905 when she began dragging. While on their way to the *Themis*, the surfmen came across one of the boat s crew who had swamped in a skiff on his way to the station for help. They picked him up and, reaching the catboat, made sail and piloted her to a safe anchorage.

On the following June 2nd the American schooner *Clara E. Rogers* stranded on Nashawena Island. It was around ten o'clock at night in a thick fog. The vessel started to break up and her captain ordered an abandon ship. The lifesavers took the crew into their surfboat and brought them to the Cuttyhunk Station. The *Clara E. Rogers* went to pieces on the rocks and was a total loss.

In a strong northwest gale in winter the big schooner-barge *Woodbury* stranded on the western end of Pasque Island. She had broken loose from her tow in Vineyard Sound. At four o'clock in the afternoon of January 9th, 1907 the keeper received a telephone call that a barge was ashore two islands to the east of Cuttyhunk. Within ten minutes the surfmen had launched and with the keeper at the steering oar they hoisted a reefed sail and moved smartly before the wind. Reaching the barge, they hailed its skipper who told them they would spend the night on their vessel.

The wind was too high to hold the surfboat alongside during the night, and it was impossible to row into that gale to return to station. So the keeper went searching for a safe place to land along that rocky coast. After two and a half hours of hard pulling they passed a small beach on Nashawena Island where they pulled their boat above the tide and found a home owned by a Mr. Manley who fed the keeper and his crew and gave them a night's lodging.

The keeper telephoned for the Revenue Cutter, *Dexter* at Woods Hole and the next morning, January 10th, the surfboat returned to the barge and sounded out the best water around her. About eight o'clock that morning a tug arrived, but had no running line to carry a hawser. The tug hauled the surfboat back to the station for an old spare whipline. Arriving back at the barge the tug's hawser was run by the lifesavers to the barge and at 3:30 that afternoon at high water the tug and the Revenue Cutter *Dexter* floated the *Woodbury* off and into deep water. The tug took the barge under tow and the *Dexter* pulled the surfboat and her crew back to their station.

Perhaps the busiest night for the Cuttyhunk surfmen was one February when the big coal schooner *Gardner Deering* of Portland, Maine ran aground in a thick fog. She struck close to shore and the lifesavers built a bonfire to see by and also tell those on the vessel they had been seen and help was on the way. The beach apparatus was set up and all thirteen men were brought ashore in safety. It had been a busy night and it seemed to be over when fresh signals of distress were observed. Another schooner, the *George Francis,* had gone aground near the same place. Again the Lyle gun was used to fire a line after which the breeches buoy was employed rescuing all five men on board. That night in heavy fog eighteen men were brought ashore from two shipwrecked vessels.

The Sow and Pigs Reef southwest of Cuttyhunk Island was a nemesis for many vessels and it claimed another victim on a night in January, 1911 when the barkentine *Stephen G. Hart* stranded in a fog. The captain and his crew rowed to the *Vineyard Sound Lightship* where they spent the night. The next morning three boats went out to the wreck. Besides the Life-Saving crew under keeper Darius Weekes and two power launches: Frank Veeder and his son, Carlton, in their boat, the *Frolic*, and Samuel Jackson piloting the *Waif*.

In heavy fog on January 28, 1911 the barkentine *Stephen G. Hart* stranded on the notorious Sow and Pigs Reef.

Vessels heading east see the first of many Cape Cod lightships, the *Vineyard Sound* #73. In the hurricane of September 15, 1944 she was blown off course and sunk, all on board died.

The surfboat had the crew of the *Hart* and a full crew of eight from the station, making a total of eighteen men aboard. That's a heavy load for that size craft and when she got hit by a huge sea, the surfboat rolled over, throwing all of them into the rough water. Frank Veeder picked up six of the swimmers and brought them ashore, while Samuel Jackson pulled the others out of the water and to safety. After the three boats had sailed out to the scene, the wind rose to a howling pitch and the waves rose higher in rolling toward the shore. The weather turned into a full gale. There were at one time eighteen men in great danger of drowning. All of them survived in a major rescue operation at Cuttyhunk.

Late in August of 1924 probably the most famous Cuttyhunk shipwreck occurred when a gale came up and forced the whaling bark *Wanderer* to destruction after she dragged anchors. The voyage was to be her last. The captain, Antone Edwards, had anchored off Cuttyhunk and gone back to New Bedford for more crew. The *Wanderer* was rolling and plunging in the wild sea as the wind increased to full gale force. When one anchor chain broke and the bark began drifting toward Sow and Pigs Reef, the crew abandoned ship in two whaleboats.

The *Wanderer*, the last whaling boat from New Bedford, America's largest whaling port, was wrecked in a gale on Cuttyhunk Island, October 26, 1924. Her crew left in two whaleboats, the first reached the *Vineyard Sound Lightship*, the other landed on Cuttyhunk.

The *Wanderer* first struck on Sow and Pigs and then was pushed by the towering combers ashore among the boulders off the west end of Cuttyhunk about one-hundred feet from the shore.

To the men of Cuttyhunk this was an old emergency that they and others before them had answered. A retired lifesaver, Louis Ramos, had a thirty-foot launch. He was joined by four islanders with long experience around small boats. They were Isaiah Tilton, Howard Cornell, Humphrey Jamieson, and John Cornell. Once out on the water, the rescuers lost sight of them in the plunging seas. One of the whale boats had made it ashore on Cuttyhunk, while the other reached the *Vineyard Sound Lightship*.

For their efforts each of them was given a gold Congressional Life-Saving medal.

Nine years earlier-in 1915- the United States Life-Saving Service and the Revenue Cutter Service were merged into the United States Coast Guard. All Coast Guard vessels after that were called cutters.

In 1952 the Coast Guard Station at Cuttyhunk was placed on barges and floated across Vineyard Sound to Menemsha. A new station was built in its place at Cuttyhunk. That station was closed on March 2, 1964.

Today there are large Coast Guard stations at Woods Hole and Nantucket Harbor and many of the sea rescues are by helicopter flying from Cape Cod.

An unusual improvisation for a Hunt gun or a Lyle gun to throw a line to a shipwreck happened on Cuttyhunk to make lifesaving history.

It was a frightening night with a full gale when the schooner, *G.W.Kimball* was driven ashore on the rocks of Cuttyhunk, March 20, 1873, with the wind from the southeast. Captain Frederick S. Allen was a volunteer lifesaver and in charge of the two stations on Cuttyhunk of the Massachusetts Humane Society. The alarm, "ship ashore," sounded and all hands started for the scene. Allen told his brother Holden to harness the horse and bring a tar barrel that was partly filled with tar that had drifted ashore and place it abreast the wreck. Then take some wreck wood and make a fire so they could see to work. When the fire was burning well the whole scene lit up showing great waves breaking over the schooner.

There was no lifesaving gear on Cuttyhunk, excepting two lifeboats, one at Station No. 44, the other at No. 42 at the east end. The east end boat was the one used to take off the crew. She was brought abreast the wreck by hauling her some 200 yards over the rocks. "When I got down abreast the wreck my brother Holden was there and the tar barrel was burning brightly and we could plainly see the crew in the rigging," said Captain Allen.

"It was too rough to launch the lifeboat as the *Kimball* must soon break up from the seas in that heavy roll from the southeast. The vessel was about 150 tons and she lay wedged among the rocks about 75 yards from the shore. The wreck about a mile from the village was not discovered until about 10 o'clock at night and it was getting along toward midnight when we got abreast of her and ready to help.

"I had an old army rifle with a steel ramrod with me and a 'quail' twine and I suggested we shoot the line over the vessel. Never before had such a thing been tried on Cuttyhunk. I believed it could be done and I would try. I made one end of the line fast to the small end of the ramrod and placed a small charge of powder in the rifle with a good wad, then I put the big end of the ramrod into the gun and let it bear well down on the wad.

"I held the gun to my shoulder and aiming the gun over the wreck, I fired. The line missed the wreck on the first shot and again on the second. However, the third shot carried the line over the arm of a crew member who made it fast to the rigging.

"The vessel's crew then pulled off a heavier one and made it fast in the rigging. The shore end of the line led through beckets at each end of the life boat, then it was hauled taut and fastened to a big boulder on the shore.

"We then made another rope fast to the lifeboat so that if the sea rolled her over the lifeboat crew would have a life line to cling to, and the men on shore could haul the upset boat and crew ashore.

"Everything made ready, Charles Church, Orrin Keeney, Henry J. Allen and F. A. Church got into the lifeboat and hauled it through the surf into deeper water where they pulled the lifeboat to the wrecked *Kimball,* and brought the crew ashore."

The Humane Society awarded the lifeboat crew $10. each for their work, but neither Captain Allen nor his brother Holden got any reward or even a mention. Allen later wrote that all the facts were not reported to the Humane Society, and he felt resentful of being ignored.

The crew of the schooner was taken to the homes of the volunteers where they were cared for until they left Cuttyhunk. The army rifle that Allen had used to fire the line was placed in a New Bedford museum.

PLUM ISLAND

The first enclosed lighthouse in America was on Little Brewster Island opposite Point Allerton at Hull, Massachusetts. It showed the way for vessels entering and leaving Boston Harbor. Boston Light was lit for the first time at sunset on September 14, 1716.

Slowly the number of lighthouses increased along the coast until the first Newburyport Harbor Light was built on Plum Island after the American Revolution in 1788. It came about because of the large amount of ocean commerce sailing the Merrimack River with its dangerous bar off the entrance. Plum Island itself and Salisbury Beach constituted a treacherous coastline. Newburyport area merchants and shipping men petitioned the General Court of Massachusetts to authorize building of lighthouses to help mariners. Until the channel of the Merrimack was stabilized with the building of jetties on both sides of the entrance, two lights were built on movable foundations as range lights with which vessels could line up a safe course to enter through the sand bars.

In 1790 the state ceded the lights to the Federal Government. There was a keeper's house nearby and a signal tower at which a white flag with a black ball in the center could be flown to notify people upriver that a vessel was in trouble off the coast. In bad weather with poor visibility or after dark, the keeper fired a cannon placed near the light by the Merrimack Humane Society. This was a practice at many lighthouses, i.e., a cannon that could be fired in fog to warn ships or as a means of alerting people that there had been a wreck. In those days the lighthouse keepers watched the sea for vessels in trouble along with their assigned duty of lighting the lamp from sunset to sunrise. In addition to watching vessels within view the assisting of ships in trouble was a voluntary community effort. River pilots and coastal residents, of whom there were not many, rounded up volunteers for a rescue and afterward helped to care for the survivors of shipwrecks.

The Merrimack Humane Society which was created in 1802 followed the practice of the older Massachusetts Humane Society in building huts of refuge along the shore offering shelter for shipwrecked people who had made it to shore.

They created small one-room buildings that contained dry clothing and wood for a fire in a fireplace. It was a welcome sanctuary out of the weather. On that bleak coast in winter it meant for those who could reach them one thing: survival.

The Federal Government appropriated money in 1848 to build and equip stations first along the New Jersey shore and then the south shore of Long Island, both areas passed by ocean-going ships heading for New York. These stations were unmanned and the actual life-saving depended on volunteers from that area. On many of these barren coastal lands there were few people living. It was not an effective system, in fact, it rarely worked. The stations deteriorated, equipment was stolen and the death toll from shipwrecks continued to read like horror stories.

On December 1, 1874 the first Plum Island Station was opened and staffed. Captain Robert Floyd was the keeper the first two years at the new station which was built close to the water at High Sandy a little more than two miles south of the entrance to the

Merrimack River. After one major storm that ate away the beach high water was six feet away and the station had to be moved farther back.

There were two problems with the location of that first station: it was too far south from the mouth of the Merrimack with its bar off shore, where so many wrecks happened. Secondly, in stormy weather the surfboat had to be launched directly into a heavy sea, a most difficult and risky operation.

This was the first United States Life-Saving Station on Plum Island. It was built in 1874 and was a standard design for the Massachusetts coast. The lookout was completely open and was reached by means of a ladder inside the second floor. A scheduled watch from up there on a cold, windy day in winter could be a grim experience. This station is listed as #1 on the map.

The next keeper was Francis L. Coffin who was appointed in the spring of 1877. He was on the job only a few days when he had his first emergency. One morning after several storms two fishing schooners came down the river and grounded on sand bars that had been created by the foul weather. One schooner, the *Flying Fish*, was quickly abandoned by her crew as waves were breaking over her. They took to several dories as their vessel filled with water.

The other schooner, the *Queen of the Bay*, capsized after striking a bar. Her crew left the ship and were in the water swimming where they were quickly rescued by the dory men from the *Flying Fish*. The plight of the fishermen was seen by a surfman in the tower on lookout. Captain Coffin called out his crew and they launched and started pulling the several miles toward the river mouth. When they arrived the men were already ashore.

Two years after he became keeper Francis Coffin resigned and was replaced by Edwin F. Hunt who became the subject of an investigation by the service. The nature of the charges against Hunt were not revealed. Whatever they were, he and his men were cleared of any wrong doing. Still there was one problem that could not be resolved: the requirement that a keeper be an experienced shipmaster. Hunt was not and so he was forced to resign. The service must have known that when he was appointed, but it took those undisclosed charges to bring about his dismissal.

In November, 1879, a new keeper was appointed, Captain James W. Elliott. A few days later on the 19th, a schooner, the *Greyhound,* went aground near the Newburyport lighthouse. The surfmen rushed to her to help get her off. They could not; at low water she rolled on her side. Further rescue attempts were thwarted by the high seas that fairly boiled into the river from the strong northeast winds. The wreck was too near the side of the channel and a few weeks later she was blown up as a menace to navigation.

Less than two weeks later a Maine lumber schooner from Bucksport went aground on the Merrimack River bar at night. The lighthouse keeper saw the *William Carroll* at dawn and raised a flag, warning the lifesavers that there was a vessel in distress.

A northeast wind was tossing high breakers onto the beach and into that sea the surfmen could not launch. They were forced to drag it north along the sands for a river launch in the basin back of the beach.

Meanwhile three volunteers manned a dory from the Point and took the crew of the schooner ashore and out of danger. There again were the twin problems with the Sandy Beach location: too far south of the mouth of the Merrimack where most of the wrecks occurred and in too exposed a location for surfboat launching in severe weather.

The report from Captain Elliott on the *William Carroll* brought swift action from the service. Within a few weeks a revenue cutter, the *Gallatin*, brought a new self-righting surfboat for the station. It was placed in a boathouse at the Plum Island Basin where the lifesavers headed to launch for the *William Carroll* wreck. Government seemed more responsive in those days and not slowed by bureaucracy, of which the Life-Saving Service had little.

Elliott had a brief tenure at the station of one year. In March, 1880 the Army Corps of Engineers began to look at the possibility of building stone jetties out from the beach at the north and south sides of the river entrance. The purpose of that was to stop the drifting of sand into the channel and make it more permanent and stable. The constant shifting of bars after each storm made it a hazard for navigation. Work began on the jetties the next year and continued for two decades.

Moving the station from Sandy Beach north was a growing priority and a number of local interests joined in an effort to persuade the service to bring that about. In a letter to Sumner Kimball, the Merrimack Humane Society, officers of the Newburyport Marine Society, the collector of customs, river pilots and local shipmasters asked that the station be moved close to the mouth of the river. There a lookout could easily observe ships in the river and approaching the coast and could help stricken vessels quickly.

Kimball agreed and the station was moved in December, 1881.

In the mean time, Elliott resigned as keeper to go back to the sea as a captain. His Number 1 surfman, George Woodman, was promoted to keeper. Woodman stayed less than two years and then resigned to enter the coastal steamship business as a captain.

For quicker launches of the surfboat this boathouse was built closer to the water with its own weathervane to the left.

After a ten year period the station was well established and back came Captain James W. Elliott as keeper. He was 60 years old, an age usually considered too old for the service except he had been a keeper earlier, so the rules were waived.

At the river mouth and off the bar and down the length of Plum Island a variety of ships, mostly schooners, stranded in bad weather year after year giving that exposed shoreline a dangerous reputation and the lifesavers many harrowing opportunities to test their skills. Northeast storms that breed off Newfoundland had nothing but open ocean across the Gulf of Maine, nothing to break their intensity before the high winds and long swells pound the exposed shore of Salisbury Beach, Plum Island and down the western shore of Ipswich Bay to Essex and Annisquam.

One October morning in 1885 the schooner *Virginia* stranded on Knobbs Beach five miles south of the station. She was heading for Rockport to load granite. During the night the vessel passed Thatcher's Island and the drunken captain gave orders to sail northwest, which passed Rockport and Halibut Point across Ipswich Bay and onto the beach at Plum Island. It was 4:30 a.m. when she went aground. The lifesavers rowed south to her and took the intoxicated captain ashore after he took still another snort of whiskey of which he had an ample supply. The schooner's crew had gone ashore in their boat earlier.

Four months later the schooner *Lizzie H. Haskell* grounded on Plum Island in heavy fog. The ship was sailing from Georgia to Boston. When the captain discovered he was aground on Plum Island he was totally surprised as he thought when passing Thatcher's Island and hearing the steam whistle that he was off Highland Light, Cape Cod. At that point he set a course for Boston. In the dense fog his ship struck the sands of Plum Island.

On January 10, 1889 at the request of the captain of the schooner *Eastern Light* three surfmen rowed to his vessel anchored off Plum Island Point in heavy weather. She was dangerously close to the beach. The three helped lift anchors and sail into the river and up to Newburyport where she was headed.

One month later the patrol south discovered a well-preserved human skeleton near the beach. Recent heavy winds shifted the sand and exposed it. There was no identification, only a pair of heavy boots typical of those worn by coastal fishermen. The station crew buried the skeleton in a deep grave back of the beach.

After a routine summer of mild weather another emergency brought out the surfmen in their boat. The schooner *Hannah Stone* was on her way up river to Newburyport on December 5 when she ran aground on the north breaker. The surfman on the north patrol saw her and fired a Coston light to show those on board they had been seen. When he received no reply he fired another and saw a lantern swinging on the deck. He then ran back and gave the alarm at the station where they launched and rowed to the scene. The surf was breaking over the stern and it was impossible to go alongside. The lifesavers backed down under the bow and threw a line aboard down which the five crewmen slid into the surfboat. The crew pulled into the river mouth and to an anchored schooner the men wanted to go aboard. The next morning the *Hannah Stone* had vanished, probably driven out to sea where she sank.

Because of the frequency of wrecks nearby the Service built its second station at Knobbs Beach in 1891. It was five miles south of the #1 station, see map.

In 1891 the United States Life-Saving Service created a second station. This one at Knobbs Beach about five miles south of the first station, which had been renamed Merrimack River Station.

The Number 1 surfman at the north station, Frank E. Stevens, was chosen to be keeper of the Knobbs Beach Station. In fact he helped choose the new location.

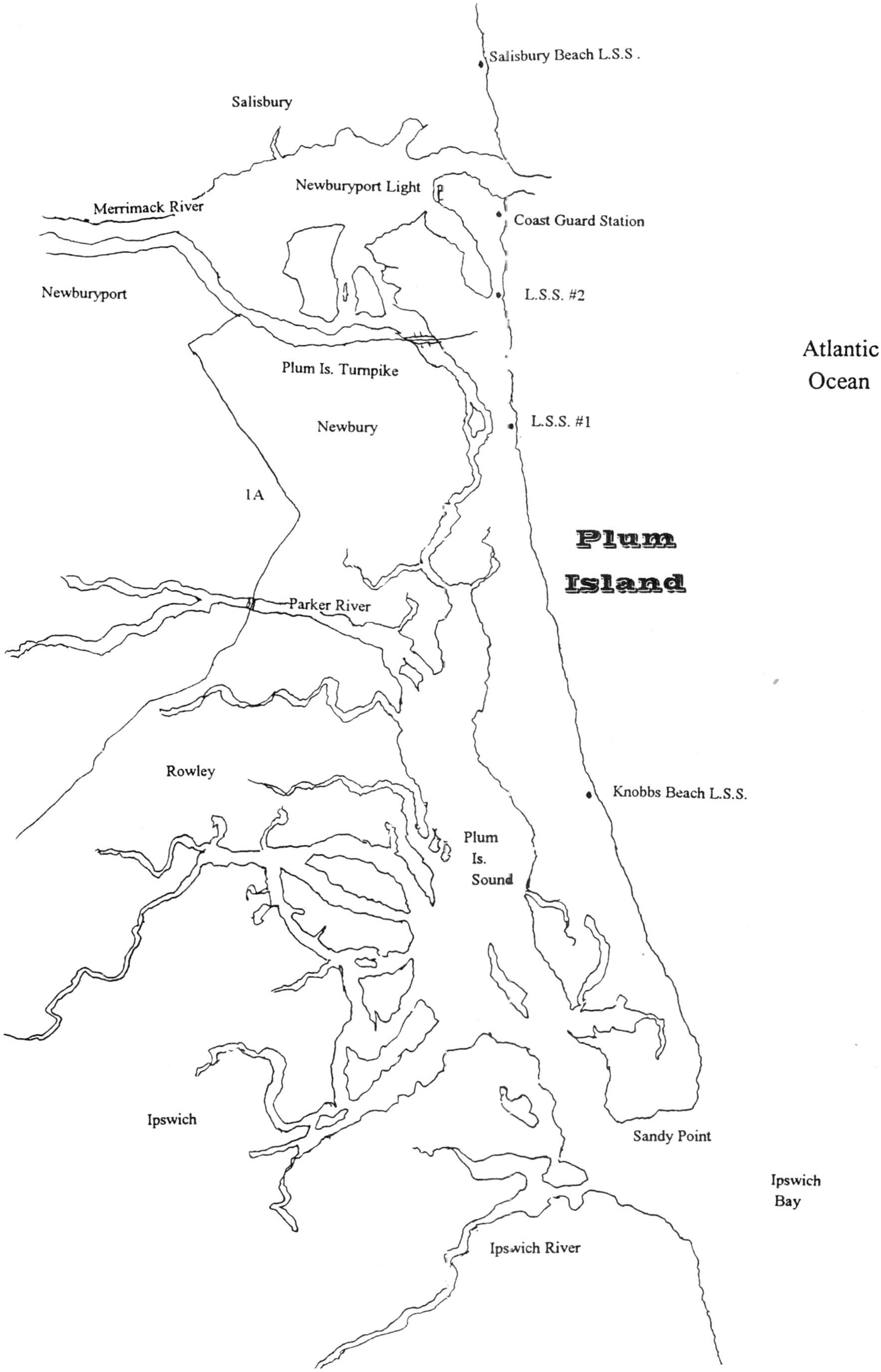
Salisbury Beach L.S.S.
Salisbury
Newburyport Light
Merrimack River
Coast Guard Station
Newburyport
L.S.S. #2
Atlantic
Ocean
Plum Is. Turnpike
Newbury
L.S.S. #1
1A
Plum
Island
Parker River
Rowley
Knobbs Beach L.S.S.
Plum
Is.
Sound
Ipswich
Sandy Point
Ipswich
Bay
Ipswich River

Stevens was a local man born in Newburyport in 1857. He went to sea with his father, a master mariner, at age 15. He was a deep-water sailor for ten years. In 1883 he joined the Life-Saving Service at 26. His assignment was as surfman at the first Plum Island Station where he served for eight years, five of which as No. 1. Stevens was unique in Plum Island lifesaving history; he was the keeper at Knobbs Beach Station for twenty-one years. During his experience as keeper he was in charge at 54 wrecks and saved 226 lives. In his long years of service Stevens suffered many hardships and privations in attending wrecks along that storm-beaten coast, on some occasions the thermometer dropping to 20 below zero. The Knobbs Beach Station was very isolated, being located at the south end of Plum Island, eight miles from Newburyport. Stevens had a coast line of five and a quarter miles to patrol; often the surfmen found it very hard to cover their patrols, especially at night, when the snow was deep and the surf ran high. His home was in Newburyport where his wife and two sons and a daughter lived. They often joined him at the station in the inactive season.

On November 2 the American schooner *Abbie S. Walker* was off Knobbs Beach with a signal in her rigging, asking for a pilot. The *Walker's* captain was unfamiliar with local channels. The crew from the Knobbs Beach Station rowed out and guided the vessel to safety.

Three weeks after that a sailboat was dismasted and swamped by breakers. The surfmen brought her out of the wild water and made necessary repairs. They carried the boat across the dunes to the Plum Island River back of the beach and launched her in the smooth water for her owners.

A few days after that the schooner *M. L. Wetherell* stranded near the Merrimack River Station. The surfmen rowed out and got her off and into deep water where they sailed her to a safe anchorage. They left the vessel only after her captain assured them she was out of danger.

One month after the New Year the American schooner, *W. H. Y. Hackett*, stranded on rocks near the Knobbs Beach Station. It was during a northeast storm with seas breaking over her and a flag of distress in her rigging.

The lookout saw her and his alarm brought out the lifesavers who launched and rescued the crew and brought them to the station for shelter. When the foul weather subsided, the surfmen returned to salvage the sails, gear and rigging. That material was given to the owner after he had been notified by the captain of the wreck.

One of the cardinal rules of the service was that no one in the service could benefit financially from proceeds of a wreck. This was absolutely prohibited.

That September a sloop yacht, the *Wizard*, capsized a mile and a half from the Merrimack River Station. The surfboat reached the scene and rescued the two men in the water. The lifesavers righted the sloop, bailed her out, reefed the sails and saw the men heading for the river and safety.

In December the schooner *A. Baker* stranded in a dangerous situation south of the Knobbs Beach Station. She had been badly damaged by drifting ice. The surfmen helped get her off and piloted her around the southern tip of Plum Island where, at her skipper's request, they ran her aground on mud flats at the mouth of a creek in Plum Island Sound. She would remain there until spring, a safe place.

The keeper at Knobbs Beach was Frank E. Stevens. Born in Newburyport, he was a surfman for eight years before becoming the headman at the new station where he served for a record twenty-one years.

The schooner *Jennie Carter* stranded at night during a storm on Salisbury Beach in April, 1894.

In the annual report of the Life-Saving Service the prior ten years had recorded twenty-five strandings at Newburyport Bar and fifteen at Plum Island. The two locations were among the heaviest for groundings in Massachusetts.

During the inactive season on May 4, 1893 four men died when the schooner *Brave* ran into a northeast gale which drove her toward Plum Island near the Knobbs Beach Station. On the schooner there were three men in the rigging and one on deck. She had sailed from Rockland, Maine with granite for Boston.

The surf was high and keeper Frank Stevens had a friend visiting overnight. This was permitted during the off season of May and June.

Stevens saw the troubled vessel and he and his friend ran up the beach where they encountered two men they asked to help. Before the four could do anything the masts collapsed, the ship turned on her side and sank. All four men drowned. The weather was described as horrible.

The next April the American schooner *Jennie M. Carter* stranded on Salisbury Beach about three and one-half miles north of the Merrimack River Station. The *Carter* came ashore during the night in stormy weather. When she was discovered the next morning the surfmen reached her and found no one aboard. They managed to save a few articles, barometer, quadrant and a few others. These they took ashore to the station. The next day with high seas breaking over her, the vessel went to pieces. Her yawl was found near the Knobbs Beach Station and surfmen from there hauled it out.

The search for bodies of the crew continued for several days; they found three which were turned over to the coroner at Newburyport.

In 1887 a man joined the United States Life-Saving Service and was assigned to the North Scituate Station where he was a surfman for ten years. He was Thomas J. Maddock who, ten years later, was promoted to keeper at the Merrimack River Station.

The new keeper at the station was Thomas J. Maddock, who had been a surfman for ten years at the North Scituate Station.

The same year Maddock arrived four more Massachusetts life-saving stations were built and staffed. The Salisbury Beach Station was two-thirds of a mile south of the state line with New Hampshire. It was close enough to be a companion to the stations on Plum Island. Salisbury Beach was the northern-most station in Massachusetts. The other three included the unique floating station at Dorchester Bay, Boston where for years there had been numerous seasonal accidents involving yachts, sailboats and swimmers.

The other two new stations in Massachusetts in 1897 were Wood End at Provincetown and Old Harbor at Chatham.

A heavy northeast wind dismasted a sloop yacht, the *Wayward,* off the mouth of the Merrimack River on August 11th, 1898 about one mile off shore. The yacht was bound to Boston from Portsmouth, New Hampshire. A short while after the yacht left Portsmouth the wind began to pick up and by the time she was off Salisbury Beach the

The station at Salisbury Beach was built and staffed in 1897. It was one of three new stations built in Massachusetts. The tower at right was an additional lookout.

In 1897 the old Plum Island 1874 station was replaced by a new building. The old structure was used as a boathouse. The new station featured a closed watch tower and larger quarters.

increasing wind created a heavy chop sea. Her captain and owner decided to duck into the Merrimack River for safety. In coming up to the wind her mast broke and went over the side, carrying the rigging and sails.

It could not be cut away as the rigging was wire and on board there were no cutting tools. Both anchors were let go, but they did not hold in that wind. She was, in fact, dragging steadily toward the high surf on the beach.

The lookout in the Merrimack River Station tower was following the drama carefully. The keeper telephoned a towboat company in Newburyport then launched a surfboat to pull for the yacht. After an hour's pull they reached the yacht and gave each of the eleven crew members a cork life jacket.

By this time the steam tug could be seen leaving the river. When the tug arrived it refused to get close to the drifting yacht to tow it into Newburyport, because of the high wind and heavy seas. The tug captain said he would land the yacht's crew if the surfmen could get them to him in the surfboat.

So the lifesavers made two trips to do that, because the yacht's crew were frightened and seasick. The tug took the surfboat in tow and steamed for the mouth of the river where the surfmen put a drag astern to keep them from being hove against the tug's stern. The wind kept pushing the dismasted yacht toward the beach and she grounded two miles south of the Merrimack River Station. The Knobbs Beach crew were watching and they and the other lifesavers went aboard and saved all the personal property and posted a guard to watch the vessel.

The following day they ran out an anchor and when the wind subsided they kedged her off the beach at high water. The two life saving crews assisted in taking her to Newburyport for repairs.

Three days later the sloop yacht, *Pert*, was anchored off Salisbury Beach. Two men tried to land through the surf in her tender when it capsized. They were thrown out, but managed to reach shore. Surfmen from the Salisbury Beach Station recovered the tender and bailed it out. The two got in the tender, and at their request, were shoved through the surf safely to the sloop.

Ten days later the keeper of the Merrimack River Station was in Newburyport on official business when he heard that a man had drowned near a wharf when swimming. The body had not been recovered. The keeper got a grapnel and with the aid of two volunteers dragged for the body without success. The keeper dived into the river several times but could not locate the body as the water was deep and muddy. Later a boat came by with a grapnel with fish hooks and the man's body was brought to the surface. The coroner took charge and notified kin.

Those events kept the lifesavers busy during the pleasant weather of August, 1898. More than three months later would come a storm that the Smithsonian in Washington would call the very worst to hit New England. A tempest that would be named for a coastal steamship that vanished at sea with everyone on board.

It was a combination of two storms that met off the coast of Cape Cod. One moving eastward from the middle west and the other heading north, creating unusually heavy gales from the southeast to the northeast. It was a monstrous storm, destructive to ships at sea and hitting the coast with a storm surge and a huge tide.

It reached Massachusetts Saturday evening, November 26 and increased in force that night and the next day when snow fell in amounts up to forty inches. Following the weekend's violence, Monday brought a drop in temperature of near 30 degrees.

They did not name storms and hurricanes then as is done today. That storm, however, ever since is called "The *Portland* Storm." Its namesake was a coastal paddlewheeler, The *S.S. Portland*, of the Portland Steam Packet Company that alternated between Portland and Boston with her sister ship, *Bay State*.

That wild Thanksgiving Weekend is often referred to as The *Portland* Gale. Why gale? In some places winds were recorded as high as 100 miles an hour, which make it a hurricane. Until the infamous October Storm of 1991 called the Perfect Storm in the book of the same name by Sebastian Junger, the *Portland* Storm had to be the most violent event over the ocean in New England weather history.

From Vineyard Haven to the waters north of Cape Ann the violence was unprecedented, according to older witnesses. The keeper at Boon Island Lighthouse off southern Maine reported wind gusts of hurricane force. North of that at Monhegan, same thing.

Mountainous combers crashed onto the beach at Plum Island and Salisbury Beach. The surfmen on beach patrol could hardly stand against the wind and flying sand. Back in town, in Newburyport the wind blew in the windows of hundreds of houses.

For all its busy maritime trade the area around Plum Island saw few ship losses during the storm. Two schooners sank. The *Newell B. Hawes* was wrecked after stranding near the mouth of the Merrimack. Lifesavers rescued the crew that Sunday, November 27. The other wind ship, the *Julia A. Decker*, grounded off Sandy Point near the Ipswich River and the entrance to Plum Island Sound. The captain was trying to reach shelter there. The other vessel destroyed was the barge *Atlantic* which foundered with a cargo of coal at Newburyport Harbor.

The only other Ipswich Bay disaster at sea was the loss of the schooner *Idella Small* at Ipswich. She stranded on Sunday, the 27th, at the height of the *Portland* Storm and then she broke up.

At the southern end of Ipswich Bay lies the village of Annisquam. The incidents that brought out the surfmen ranged from the mouth of the Essex River past Crane's Beach on the western side to Lanesville and Halibut Point on the east. The station was called Davis Neck and it was situated just north of the Annisquam Lighthouse which was on Wigwam Point. Through the years a number of schooners ran into trouble striking the bar at the entrance to the Annisquam River.

The trouble spot for the Davis Neck lifesavers was Lane's Cove at the Village of Lanesville about one and one-quarter miles northeast of the station. Vessels were sailing in and out of there most of the year to pick up granite from a large number of local quarries.

A high breakwater on both sides of the opening kept the small harbor safe from the sea. There were derricks there for loading and three main wharves where a well-known local man, William Robb Cheves, a Scotsman from Kennay, Aberdeenshire, built his granite business after the Civil War. Cheves, starting in 1894, furnished most of the stone for Dog Bar Breakwater at Eastern Point, Gloucester. The breakwater was built on a base of rubble stone of about 75-thousand tons. The flat cap over the rubble consists of five hundred blocks, ten feet wide. Cheves shipped those blocks from Lane's Cove for the capping. In addition to that Cheves and the other local granite operators produced hundreds of thousands of paving blocks which were stacked along the wharves at Lane's

Cove. There were so many vessels moving stone from Lane's Cove, Bay View and Halibut Point that for years the Massachusetts Humane Society had a Lanesville boathouse with a surfboat manned by a locally-recruited volunteer crew of lifesavers, most of them fishermen.

Lane's Cove on Ipswich Bay was an artificially-created harbor protected by a high breakwater on the ocean side. It was north of the Davis Neck Station by a mile and a quarter and was the shipping center for the local granite industry.

The Davis Neck Life-saving Station was a little more than one mile south near Annisquam Light. The crews there were responsible for marine accidents up to Ipswich on the west and to Bay View, Lanesville and Halibut Point on the eastern side of the bay. The Davis Neck Station was closed later and moved to Straitsmouth at Rockport.

The granite business on Cape Ann closed down about 1920. There were occasional shipments by water after that, mostly for government contracts for breakwaters and jetties along the New England coast.

A Newburyport jetty was completed in the late 1930s with the last granite shipped out of Lane's Cove. A steam lighter, the *Rocket*, towed scows for the Frederick Starr

Company of New York across Ipswich Bay. One October night before World War Two a scow returning empty to Lanesville developed a major leak and sank with the solitary crew member climbing to the roof of the deckhouse of his rapidly sinking vessel. Someone on the tug spotted the man and he was rescued. Then the scow was cut loose and sank in several hundred feet of water in Ipswich Bay.

The granite quarries on Cape Ann delivered their product by water. There were granite-loading piers at Pigeon Cove where the Rockport Granite Company had tracks leading back to its quarries. Also at Pigeon Cove the Pigeon Hill Granite Co. had a pier with the terminus of an inclined railroad that descended from the company's two

William Robb Cheves, wearing straw hat, is in the middle of his granite business. This is Deep Hole in Lanesville the source of much of the stone for Dog Bar Breakwater in the entrance to Gloucester Harbor. Cheves produced most of the stone for it.

quarries in the hills back of town. The railroad was a single-track affair that ran on gravity down to the water. A team of horses pulled the empty flat cars back to the quarries for loading.

At Bay View on Ipswich Bay the Rockport Granite Co. had a long wharf extending into the harbor where derricks loaded stone. Tracks led back up in the hills to the quarries that were originally the Cape Ann Granite Co. until 1894 when the company was sold to the growing Rockport Granite outfit. Steam locomotives moved flat cars from the quarries to the wharf at Bay View.

Next to Lanesville was Bay View where the Rockport Granite Company operated Old Pit back of town. These rails ran into the hills where steam trains loaded granite to ship from this wharf. The lighter is loading a column for Boston. Davis Neck Life-Saving Station was just south of here on Ipswich Bay.

At the entrance to Ipswich Bay The Babson Farm Quarry was at Halibut Point where the old wharf is now a parking lot with a restaurant. The cove there was deep enough for big schooners to load. Today there are a few lobster boats at anchor. The ships ranged from granite sloops to schooners of all sizes: with two, three and four masts. Steam lighters from Rockport and Boston usually took deck cargo of paving stones and sometimes large pieces of granite. Granite blocks and columns for public buildings-court houses, post offices-were usually carried on schooners and lighters.

Lanesville had three granite piers with the longest in back of a breakwater that protected the harbor from Ipswich Bay. Ships entered through an opening called the gap, the other side of which had a shorter pier also protected by a breakwater known as the wall.

Directly across the cove from the gap was a large granite pier with derricks for loading vessels. Teams of oxen were always present and used to haul large wheel carts from the quarries with the stone which could be any of the local product: edgestone, foundation blocks, capping, wharfstone and doorstones for steps to houses, barns and other small structures. Paving blocks were piled everywhere on all the piers and went to road departments in cities and towns on one vessel after another out the gap, down Ipswich Bay and out into the Atlantic.

The granite ships took up most of the space in the small harbor. The other vessels were owned by fisherman whose gear was stored in wooden shacks above the beach on the north side also protected by the long breakwater. Lane's Cove was a busy place. There was also a Massachusetts Humane Society boathouse with a Society surfboat. A local volunteer crew was composed of a keeper and surfmen, most of whom were fishermen.

The danger nearby was Lane's Point about a thousand feet north of the cove. The point is granite down into the water and open to the seas coming from the northeast where violent storms are born. The waves that crash onto Lane's Point break into spray the wind often carries hundreds of feet south toward the cove. During such weather the whole of Ipswich Bay is white water churning. It is a frightening sight.

In one five year period toward the late 1880s there were six strandings on Lane's Point with each vessel a total loss. They were carrying granite away from Lanesville, except one, when they were caught in bad weather and grounded.

The other schooner was carrying lumber from Bangor, Maine to Norwalk, Connecticut when she sprung a leak during a terrific northeast gale. The vessel, the *Eurotas*, became water-logged and stranded on Lane's Point where she immediately went to pieces. The five man crew was rescued by a passing schooner and landed at Lane's Cove where they were then taken to Davis Neck Station by the keeper. They were made comfortable at the station and given dry clothing provided by the Women's National Relief Association and sheltered several days. Most of the lumber was scattered over the shore and two-thirds of it was saved by wreckers.

On another occasion a schooner from Maine stranded on Lane's Point with waves breaking over her and a distress flag in her rigging. The Humane Society group rescued the five man crew in dories and took them into the cove. When the surfmen from Davis Neck arrived the crew had been brought ashore and the cargo of lumber was piling up on Lane's Point. Some of it was salvaged.

One August afternoon the schooner, *Caroline Young*, of Philadelphia left Lane's Cove with granite and sailed into high northwest winds which drove her onto the point. Lifesavers from Davis Neck saw her and pulled for the scene in their surfboat.

It was a tricky situation but they managed to back down under the boom and take the crew off and land them at Lane's Cove. The *Young* was valued at $5,000. and the granite at $1,200. The loss was total. The captain should never have left that protected harbor into such a strong wind. He had little room to maneuver once he went through the gap into Ipswich Bay.

One November night in 1888 a gale struck, breaking the mooring lines on the schooner, *Venus*, of Gloucester. She was laid up for the winter in Lane's Cove and tied to a granite pier. There was no one aboard the vessel which floated free, struck rocks and sank. The Davis Neck Station was notified and they raised and hauled her onto a beach behind the breakwater where she lay until spring.

One January night in 1893 the American schooner, *E. F. Willard*, sailing in Ipswich Bay, mistook Ipswich Lighthouse for that of Cape Ann Light on Thatcher's Island, Rockport. The *Willard* stranded on the bar at night. The crew abandoned the vessel and left in four dories. The surfman on the east patrol from Davis Neck Station came across three of the men and escorted them to the station in Annisquam. The keeper sent out an extra patrol and placed lanterns on the beach to guide the others ashore.

Seven more of the schooner's crew were found later and taken to the station. That meant ten of the vessel's crew of fourteen accounted for. At the station, the crew were given food and dry clothing and otherwise cared for.

The other four crew members had rowed a dory toward the Ipswich Light and landed on the beach near there. They went to the light and spent the night and the next day they joined their mates at Davis Neck.

The following morning the lifesavers pulled out to the *Willard* and found her full of water. There was a surfboat from Knobbs Beach Station but no one from either boat could board the stricken ship as the seas were breaking over her.

The surfmen are at their oars and the keeper has just jumped into the stern of this Plum Island surfboat. Into the fury of an eastern storm these men will now try to ram the boat into that wave. Often the Atlantic Ocean threw them back many times before a successful launch. It was one of the hazards.

Three days later the wind had calmed down and the surfmen helped to pump her out and get a line to a tug which towed the schooner to Portland, Maine for repairs. The fourteen men in her crew stayed two days and then went to Gloucester for a train to their homes.

One beautiful August morning the keeper of the Davis Neck Station was out in his skiff at eight o'clock, enjoying a bright day and a gentle breeze. In the vicinity was a small boat, carrying five men. Without a warning the boat capsized through carelessness and all five men wound up in the water. The keeper and another man who was nearby in a dory pulled quickly to the rescue. Two of the men were helped into the dory, one into the skiff, and one, being too large and heavy to get into the skiff, was supported in the water while the keeper rowed to the shore and landed him. The accident happened only a short distance from the beach. The fifth man saved himself by swimming to the shore.

The men were on a drunken spree, and had stolen the boat in which they came to grief. Their boat was recovered by the keeper and restored to its rightful owner.

The schooner, *Flying Eagle*, of Portsmouth, New Hampshire was on its way to Boston when the captain decided to stand into the Annisquam River for a harbor until morning. It was 2 a.m. and the wind was fresh from the northeast when the ship went westward of the channel and ran aground.

The nearest help was the Davis Neck Station to the east from which the vessel could not be seen. The information that there was a ship ashore did not reach the station for another twelve hours.

When they heard, the surfmen launched and pulled for the vessel, finding her high and dry. They took out an anchor and tried to kedge her off the beach, but without success. After the tide turned, the work was discontinued and the lifesavers returned to the station.

At 3 o'clock in the morning they went again and renewed their efforts, but the tide was not high enough to float her. On the next high water the surfmen returned to the schooner and succeeded in releasing her from the sand. They sailed her to Bay View where they made her fast to the granite pier. The *Flying Eagle* sustained only slight damage. Her two-man crew was sheltered during those days at the station.

At the southern tip of Plum Island the steep beach was a popular place for bathing and picnicking. It was called Sandy Point. The volume of fine beach sand also made it an ideal location for schooners known locally as sand droghers. These vessels, usually two-masted schooners from the fishing business, were too old to be sailed in rough water far from shore. They carried away sand for the building trade, usually to Boston.

The droghers anchored close to shore at high tide in such a way that as the tide ebbed the vessel would lean toward the beach. The crews varied in number from three to five. Planks would be laid as a walkway for the men to push wheelbarrows up to the schooner from the nearby digging site. The owners of the beach were paid for the number of tons removed. These vessels were a familiar sight at the Point and at the extreme northern end near the mouth of the Merrimack River where the sand had a coarser quality. In either place the work of loading a drogher was strenuous. Each man was expected to load fifty tons in two tides.

On April Fool's Day, 1900, the American schooner, *Minnesota*, stranded while anchored near the shore in preparation to load sand. It was low tide and the seas were rough. The keeper at the Merrimack River Station telephoned for a tug and then boarded the vessel with his crew, but nothing could be done until high tide.

The tug arrived and the surfmen ran a hawser to the schooner, as a sand bar prevented the tug from approaching near enough to put a line on board.

Surfmen yanked on anchor lines while the tug pulled and she still stuck fast. The line parted at one point and the lifesavers ran another. No luck. The tug gave it up for the day and returned to Newburyport.

The following morning the lifesavers laid out the anchors again, but still could not float her. A heavy sea started to roll in and the schooner was rocking in place and then began to break up. The surfmen saved the three anchors for her owners and the *Minnesota* became a total loss.

Beaching a surfboat after a rescue could be a dangerous business, too. These men can relax now that they are on land. The continuous practice with all their equipment made them confident.

These Salisbury Beach lifesavers often teamed up with the Plum Island crews for a double effort at a rescue. It was one of the advantages of placing stations fairly close to each other.

In September of the following year another sand drogher came to grief when she sprang a leak and sank two miles north of the entrance to Merrimack River. The lifesaving crews from the Plum Island Station and Salisbury Beach rowed to the schooner, *Julia A. Decker*, but after throwing thirty tons of sand overboard were unable to float her.

The rest of her cargo shifted forward and buried her head under water. The lifesavers tried to operate the pumps, but it was impossible to free her. The skipper decided to dismantle his ship. Five days later the crews from Plum Island and Salisbury Beach went on board and assisted in taking her sails, rigging, furniture and blocks to the shore. At the next storm the *Julia Decker* went to pieces.

Three months later almost to the day during a thick snowstorm the surfman on day patrol saw a man in a dory rowing for the beach. He waded into the water and assisted the man to reach shore without capsizing. The lifesaver hauled the dory above the high water mark and escorted the man to the station. The keeper sheltered the man and fed him for a day, then took him to the Newburyport railroad station and got him a pass to Portsmouth where he hoped to rejoin his schooner, the *Elsie F. Rowe*. The man was dory fishing and had gone out of sight of the vessel during the storm.

On the same day an anchor chain parted on the American schooner, *Onward*, pressured by heavy drift ice in the Merrimac River, causing the vessel to ground on a reef one mile north of the Plum Island Station. It was 7:30 in the evening of December 17, 1901. The keeper spotted the ship from the tower and telephoned a towing company in Newburyport to send a tug. Too much ice in the river, so the company refused.

In order to help the men on the schooner, the keeper launched the surfboat but the men could not row through the packed ice on the south shore. It was a barrier. After staving a hole in the bow, the keeper gave up that idea.

Next, Captain Thomas Maddock telephoned the keeper at the Salisbury Beach Station to have them launch a boat to board the schooner as the north side of the river was free of ice. Maddock, meanwhile, had his surfmen haul the beach apparatus to a spot opposite the wreck where they prepared to fire a line across the vessel for a breeches buoy rescue.

The schooner, *Onward*, was 900 feet from the shore when Maddock's men fired a line. When the men on the schooner did not haul off the whip line, the Plum Island crew prepared to fire another shot. The wreck was gradually drifting closer to the beach when the Salisbury Station surfboat was seen near her stern. Those lifesavers used a heaving stick and then boarded the vessel and took the lines from the opposite shore.

The four men on the ship, one of whom was badly frostbitten, were rescued by breeches buoy, then taken to the Plum Island Station for two days shelter and food.

The Salisbury Beach crew cast off the breeches gear from the wreck and then rowed back to the north side of the Merrimack. It all happened in the middle of the night.

The next day the Plum Island surfmen boarded the schooner after the water around her became clear of ice. They threw off about 45 tons of sand before a tug came down river to haul the wreck to Newburyport.

The two lifesaving stations teamed up for a successful rescue on this occasion and on others; it was part of their mission. The placing of stations that close together was the idea of Sumner Kimball, proximity made cooperation possible.

Ice in the river caused another problem when the following February the schooner, *Eastern Light*, anchored in the Merrimack River was struck by another schooner which parted her chains from the ice accumulation. Surfmen from Plum Island went to her and parted the two vessels. They found her outer planking cut through just below the water line on the starboard bow. The *Eastern Light* was filling rapidly. To prevent her sinking in the river, the surfmen beached her and nailed tarred canvas and planks over the open seams. At high water she floated off the beach.

A pile driver dragged her anchors one afternoon that May and stranded on the shore of the Hampton River some two and one-half miles north northeast of the Salisbury Beach Station. The keeper telephoned for a Newburyport tug and piloted her into the narrow river where the line parted and the barge grounded again.

The next afternoon the surfmen helped get the vessel off and took a line to a tug which hauled the pile driver to Newburyport.

The following day the freight steamer, *Globe,* from Bath, Maine stranded on the bar at the mouth of the Merrimack on her way out to sea. The lifesavers from the Plum Island Station pulled to her in the surfboat. When she floated later on a rising tide, they acted as pilots, directing her captain where he could find the deepest water. The *Globe* made many trips along the coast. She was only two years old when she grounded that day. Built at Richmond, Maine, she was 91 feet in length, weighed 137 tons and had a crew of six.

The tugboat fleet in Newburyport did a good business, towing schooners in the Merrimack River. Vessels would fly signal flags P over N, “Want a steam tug,” when they were off Plum Island. They also required a tow on the way out.

There were so many strandings at the mouth of the Merrimack. One day the lookout in the tower at Plum Island Station saw a tug with two barges close to the beach and raised signal flags, J D, you are too close to the shore. The tug captain said "thank you" with two blasts on his steam whistle and headed for deeper water.

The schooner, *Herbert M. Rogers*, stranded on the north side of the Merrimack River on a pleasant September day in the noon hour. The keeper, Thomas Maddock, rowed over in the dory and at the request of the captain telephoned for a tug. The vessel was to arrive at the next high water. That night at 11:30 Maddock and his crew went back to the *Rogers.* The tug was not there, so the surfmen ran out anchors and tried to kedge the schooner off her grounding. She would not float.

The next day the Newburyport crew and the men of the Salisbury Beach Station rowed out to the *Rogers* in separate surfboats. They ran a hawser to a tug and after the schooner was floating, assisted to furl sails and bring her to a safe anchorage.

One pleasant afternoon in early May of 1904 the lookout at Plum Island Station saw a gasoline-powered pleasure craft with a disabled engine. The lifesavers launched and pulled the surfboat to the yacht and put a line on board and towed her into the Merrimack River and a safe place.

Two days later the barge *Hercules* stranded in the Merrimack and the crews from Newburyport and Salisbury Beach went out to the vessel and ran lines to tugs which pulled her free.

Aside from those two incidents, life at the two stations was mostly six days of drills and maintenance with Sunday for visiting when family members showed up for a picnic.

The summer with its pleasant skies was usually a time in the lifesaving field for occasionally chasing drifting skiffs or catboats-nothing major.

On Ipswich Bay the last day of the inactive season, July 31, 1905 two men went out in a launch about 6 p.m. to the sloop *Dipper* to condition the vessel to weather an approaching storm. When the sloop went adrift they boarded her and dropped two anchors in the rising sea that did not hold. So they put overboard a large piece of iron ballast which brought her up. While they were busy they did not notice where the *Dipper* was drifting and when the ballast finally held her, they were in dangerous waters about 1,500 feet off shore. They were unable to escape by their launch.

Shortly after nine o'clock, the keeper of the Annisquam Light spotted the men and telephoned the Gloucester Station. The keeper there rounded up a volunteer crew and they went to Annisquam but found it too dangerous to launch at the lighthouse where there was a Humane Society surfboat.

They engaged a wagon to haul that surfboat to the Davis Neck Life-Saving Station, taking the lighthouse keeper along to act as pilot.

The keeper from Gloucester with the keeper from Annisquam Light and the volunteers pulled out into Ipswich Bay to a place where they thought the two men were. When they did not find them, they burned a Coston signal. Getting no answer they went in closer to the rocks and burned another signal. Soon after that they heard cries for help from a breaking ledge. A rescue was made about five miles north of the station.

Within a week the Salisbury Beach lookout saw a disabled launch several miles southeast of the station about two-hundred yards off shore. The active season had just begun and the crew launched and pulled to the craft which they took in tow for Hampton River. A power boat came by and offered assistance, taking the launch in tow the rest of the way.

Although a gasoline-powered launch was warned by the signal flags, J D, "you are sailing into danger," from the Plum Island Station at Newburyport, the launch ignored the signal that the bar at the mouth of the river was impassable. They tried twice to cross it. With the second attempt the lifesavers pulled for the launch in a surfboat and found

Annisquam Light was the first of four built on Cape Ann. On the right is the Humane Society's Boathouse with a dory on the ramp. The keeper of the light was also the Society's keeper.

Keeper Maddock and his wife Florence had six children. He is shown in front of the boatroom with the four youngest. During the inactive season-June and July-the whole family lived at the station. For Tom and his family it was life at the beach.

By 1905 when this photograph was taken the Plum Island Station was already named Merrimack River Station. It's a warm day. The windows in the lookout are all open.

her sinking. All but one of the five occupants were picked up by the surfmen who had the launch in tow. The missing man was found by the station crew clinging to the bottom of a skiff that had gone adrift from the launch.

All five of the rescued people were sheltered at the station over night and given dry clothing from the store provided by the Women's National Relief Association. Their rescue took place on August 28, 1905.

Two schooners while underway drifted towards Ipswich River Bar on September 21 and stranded in a dangerous place. Half the crew from the Plum Island Station went to one of the schooners and half to the other. A heavy anchor was carried out from the schooner, *M. L. Weatherell*, and a hawser from her to the other, the *Eastern Light*. With the incoming tide, sail was made on both vessels, a strain was put on the cables, and at high water both floated free.

The Plum Island Station responded to two emergencies the morning of November12 when two men were seen in the water struggling in the channel under the bow of a schooner. They were in a dory pulling down the channel when they were run down by the schooner. The surfmen reached them in a boat and brought them back to the station, after which they left for their homes. The dory was destroyed in the collision.

A short time before, the lookout reported to the keeper that smoke could be seen from Whaleback Shoal to the north. The lifesavers pulled to the scene and found the American steamer, *R. J. Hackett*, on fire and abandoned by the crew. They boarded her and managed to save the officers' papers, several compasses, and some clothing in a trunk. These things were turned over to the owners. The *Hackett* was enveloped in flames and sank.

Lifesaving in bitterly cold weather was most difficult for surfmen in long rescues. It was like that the following February when one morning the sloop, *Fortuna,* stranded in hazy weather in the north breaker on Ipswich Bar. She was seen and reported by the keeper of the Ipswich Light.

The surfmen cleared her of ice and threw her ballast overboard. After running her anchors out, they stood by for the incoming tide. At high water she started to leak badly. The surfmen ran the pumps, but could not keep up with the water.

The lifesavers took the master and his crew to the lighthouse nearby where they were cared for. The following day the heavy surf pounded the vessel and she filled and broke up.

By 1906 the former Plum Island Station was now designated Newburyport. On April 10 the American schooner, *Marguerite*, dragged her anchors during a southeast gale in thick weather and stranded on Black Rocks one and a half miles north of the station. One hour later the keeper was called with the word:"ship ashore!"

The crew reached her and set sail after hauling anchors. Helped by a favorable wind, she floated and they guided her to a safe anchorage.

Nine days after that an accident involving three men in a fishing dory resulted in the death of one of them despite the efforts of the Newburyport and Salisbury Beach stations. It happened around midday when Thomas Maddock at the Newburyport Station noticed the men outside the mouth of the river in a choppy sea and an ebbing tide. They were obviously in trouble and tried to make a landing along side the north jetty, but without luck.

So they headed the dory toward Salisbury Beach and after half a mile turned for the shore to attempt a landing through the breakers. Their craft was lifted by a combing sea and capsized. The three momentarily clung to the overturned dory until a heavy sea swept it away. Two of the men were powerful swimmers and they reached shore alive. The third man could not swim and he was swept out by the undertow and disappeared.

When Maddock saw the three heading north he telephoned the Salisbury Beach Station and told them a dory was heading their way that needed watching. Maddock then launched his surfboat and they followed. The surfmen from Salisbury ran down the beach toward the dory which then was heading through the breakers for a possible landing. Before either group could reach the fishermen the tragedy occurred. For several days a search was made for the missing body, without success. The two swimmers were taken to Salisbury Beach Station and given dry clothing from the supply. That September the schooner, *Newell B. Hawes*, which had been in trouble at Plum Island before, was about to sail with a cargo of sand when she sprung a serious leak. The Newburyport lifesavers arrived and threw about 60 tons overboard. The surfmen ran the pumps all night and kept her afloat until morning when they beached her on the flats.

In the middle of November, 1906 the schooner, *Wm. F. Green*, struck a derelict and lost her rudder and was leaking. She anchored in a dangerous place less than a mile north of the Salisbury Beach Station with a distress signal in her rigging. She was seen by the beach patrol about 5 a.m.

Within a few minutes the surfboat was launched and pulled quickly to the scene. The lifesavers helped at the pumps while the keeper went ashore and called for a tug. The surfmen got up the anchor and worked her into deep water. The tug soon appeared and took her in tow for Portsmouth with the lifesavers staying aboard to work the pumps at the captain's request. The tug hauled the surfboat back to the station at 4 p.m.

The schooner, *Bessie C. Beach*, stranded seven and a half miles south southeast of the Knobbs Beach Station, now called Plum Island, on a bitterly cold December 8. There was a heavy fog over the water and the wreck was not visible from Knobbs Beach. The keeper there was notified by telephone. The keeper at Gloucester was notified by telegram. The Gloucester crew obtained a team of horses that hauled the surfboat overland seven miles to Annisquam where they launched into Ipswich Bay.

On arriving at the wreck, sails, rigging and some of the spars were found to have carried away. The schooner crew was badly frostbitten. The lifesavers treated the sailors for frostbite and several of them from Knobbs Beach went ashore and telephoned for a tug. The surfmen from both stations cleared up the wreckage, while the Plum Island lifesavers stayed through the night, working the pumps. The Gloucester crew returned to Annisquam and hauled their surfboat across land back to their station.

The next day a Newburyport tug arrived with a lighter. The Knobbs Beach surfmen helped lighter the cargo, ran a hawser from the tug and worked the pumps until the wreck was safely docked in harbor.

The beach patrol in winter at Salisbury Beach Station with each man carrying a time clock to record the end of the walk. At that time of year it was a lonely, isolated place.

One day in late June during the inactive season, a local catboat, the *Strideway,* stranded on a sunken wreck about a mile north of the station. Thomas Maddock, the keeper of the Merrimack River Station, and his son, Richie, hastened to the scene, finding numerous holes punched in her bottom by wreckage.

They stayed there until low water when they bailed out the boat, stuffed the holes with oakum and patched them over with canvas. They laid out an anchor and at high water hauled the catboat off, set sail and took her into Newburyport Harbor and back to her owners.

In April, 1908 the Gloucester surfmen had to combine efforts with the Plum Island lifesavers, because the Davis Neck Station in Annisquam was moved to Straitsmouth at Rockport. The Knobbs Beach Station was already renamed Plum Island. The schooner, *Gatherer*, in the lower harbor at Essex, just off Ipswich Bay, stranded. The wreck was more than seven miles south south-east of the Plum Island Station.

The keeper was notified by telephone and he then called the Gloucester Station. The Plum Island men pulled down Ipswich Bay in their surfboat and the Gloucester surfmen came in their new power surfboat. Not all stations had them yet.

Upon arrival the surfmen discovered the vessel high and dry, badly strained and leaking. They ran out an anchor, lightered part of the cargo, and on the rising tide kept her pumped out. At high water she floated and they towed her with the power boat to a safe berth in Essex Harbor.

One accident with some of the greatest loss of life at Plum Island happened May 10, 1908 about one mile off the bar at the mouth of the Merrimack River. There was a high sea, and a gasoline launch with nine occupants broached to while running before it. The pleasure boat capsized throwing everyone into the water.

The accident was discovered by the lookout in the tower of the Merrimack River Station. The surfboat was manned and started for the scene northeast of the station. Two fishing boats, which were lying off the bar awaiting a smoother sea before coming in across the bar, were the first to arrive at the accident.

They picked up four of the men found clinging to the bottom of the launch. The five others were lost in those heavy waves. Captain Maddock headed for the one fishing boat closest to the scene and picked up the four survivors; one of them was so far gone that he had to be resuscitated by the surfmen on their way ashore.

Both fishing boats started searching for bodies of the five missing men. None were found.

After a hard pull through a sea which was breaking in the channel, the lifesavers returned to the station where the rescued were put to bed after receiving medical attention. They later were given outfits from the Women's National Relief Association. One of the men was kept overnight at the station. The launch was picked up by a fishing sloop four miles south-east of the station off Plum Island and was towed to Rockport.

Three weeks after the capsizing two of the missing men's bodies were found by members of the Merrimack River Station.

Around Thanksgiving of that year the fishing schooner, *Thomas A. Cromwell*, stranded during thick fog a half mile north of the Plum Island Station at Knobbs Beach. The *Cromwell* was discovered by the beach patrol who fired a Coston flare, telling those aboard they had been seen and that help was on the way.

The crew of twenty-one did not wait for the surfboat; they came ashore in their dories and were given food and shelter overnight.

The keeper telephoned to Boston for a tug as that city was the *Cromwell's* home port. The next day the tug *Orion* arrived. Surfmen ran a hawser from the tug to the schooner, and she was hauled into deep water and towed to Boston.

Some vessels just can't seem to stay out of trouble. The *H. F. Dimock*, steamer of the New York to Boston Metropolitan Line, comes to mind immediately.

Around Plum Island the schooner *Newell B. Hawes* dragged anchors in a heavy gale and stranded more than three miles south southwest of the Plum Island Station not far from Sandy Point.

The vessel sprang a leak, filled and sank. The surfman on patrol fired a Coston flare and ran back to the station. The crew went to the scene, but the *Hawes* was beyond their help.

The keeper telephoned news of the accident to the owners, and requested that he send wreckers. Two weeks later the schooner was floated. It was April 7, 1909.

This was the third stranding on Plum Island for the *Newell B. Hawes* and her third rescue. She lived, like the *Dimock*, to sail again.

The day after Christmas, 1909, the schooner *Ada K. Damon* parted her anchor chains and stranded three and one-half miles south south-west of the Plum Island Station.

It was at night and the vessel was not seen until the next morning. The lifesavers launched and went to the wreck and found her full of water with seas breaking heavily over her. Her crew were safe on shore.

On the 28th the surfmen went again to the schooner, but her captain declined assistance, saying he was going to sell the vessel to wreckers. On January 2 the surfmen again went to the wreck, and her captain said he could find no wreckers to buy her, and asked that the lifesaving crew assist in stripping the vessel the following day. The surfmen helped, as the schooner was totally lost.

On September 17, 1910, a party of six men started out at Haverhill, Massachusetts in a flimsy 26-foot power launch for a two-day trip down the Merrimack and up the coast to Portsmouth, New Hampshire. For them it was a disaster that only two survived. On reaching the mouth of the Merrimack they found the water on the bar off the entrance so high that some of them wanted to abandon the trip because of fear. Others wanted to go on. They decided to settle the matter by having a discussion on shore.

They headed for shore next to a steamboat landing at Salisbury. Once there the group decided to go on, despite the arguments of the oldest man there who was the father of two of the others. In leaving along side a jetty, a piece of the tiller broke, putting the steering gear out of commission. Two of the men attempted to steer the launch by holding the rudder with their hands.

When the craft ran into the swell from the ocean, the boat pitched violently, throwing them into the waves. The two tried to swim to safety, but they were overcome by the huge swells and sank. The father, Charles S. Fountain, now held his other son by the legs while the boy held the rudder. The heavy swells tossed the launch so violently again that both of them were thrown out as the others had. The son disappeared, but the father managed to reach shallow water where a by-stander helped him out.

The same wave that threw the father and son out swamped the launch, her stern struck bottom and only the top of the cabin was above the water. The other two men clung to the cabin despite the combers breaking over it, rapidly exhausting their strength.

One of them decided to try for the jetty 150 feet away. A sea broke over the swimmer before he had gone far. When the wave passed he was no where to be seen.

The last survivor out in the water stayed with the half sunken launch. He was rescued soon after that by a crew from the Merrimack River Station who had watched these amateurs with concern. The surfmen launched and pulled to the scene against a strong sea, but not in time to save more than the one man. The father, Charles S. Fountain, who wanted to turn back at the mouth of the river, lost both his sons and very nearly his own life.

So many of these stories about Plum Island and Salisbury Beach through the years tell of heavy seas sweeping toward the shore and the bar off the mouth of the Merrimack River. There is nothing to break the winds from the east and the long swells of the Atlantic Ocean. Many miles of the Massachusetts coast are as exposed and vulnerable as the Plum Island area.

For example the United States Life-Saving Service always kept records of all aspects of ship wrecks. From 1902 to 1911 their annual reports show that there were 31 strandings of vessels at the Newburyport bar. At the same time Plum Island had eight. A total of 39 strandings of major vessels, during those years, the highest for any local area on the coast of Massachusetts.

A dangerous place for mariners and a busy one for lifesavers.

One year after the Life-Saving Service became the Coast Guard the surfmen had to rig a breeches buoy to save the crew of the schooner *Virginian* on Salisbury Beach.

ROCKPORT AND PIGEON COVE

Along every dangerous coast the first and the oldest system to help mariners and save their lives was the building of lighthouses. The first lighthouse in America was the Boston Light on Little Brewster Island north of the south channel.

The lifesaving system created by Sumner Kimball in 1871 was a back-up system to that much older Lighthouse Service. There were never as many life-saving stations as lighthouses. In the first years there were six life-saving stations along the entire coast of Maine. By 1887 only one more had been added. Compare that with the sixty lighthouses-or stations- as they were known along that same coast.

From Newburyport to Gloucester there were eight light stations. At Newburyport there was the Plum Island Light on the point just south of the Merrimack River near its entrance. Up the river in the inner harbor there were two light towers that were range lights to guide vessels to the port. That made three beacons at Newburyport.

Down Ipswich Bay to the south was Ipswich Light in the sand dunes, guiding vessels into Ipswich River.

On the eastern end of the bay was Annisquam Light on Wigwam Point. It was the oldest of the Cape Ann lights, dating from 1801. Near the mouth of the Annisquam River and its harbor of refuge, the Annisquam Light was located close to the southernmost extreme of Ipswich Bay opposite the white sands of Wingaersheek Beach. It is a handsome masonry structure of medium height. Within a few yards of the beacon was a Massachusetts Humane Society boathouse with a launching ramp in front.

Off Rockport on Thatcher's Island the famous Cape Ann Light Station with its twin masonry towers was built in the 1860s. Each of the beacons is 124-feet high, often referred to as the Twin Lights. In the winter of 1932 the northern light was turned off as an economy measure during the depression.

A few miles to the south at the entrance to Gloucester Harbor is Eastern Point Light, dating from 1832. It is a warning and a welcome to the most famous fishing port in America. In the 1890s the tower was rebuilt to look like a replica of Annisquam Light.

During that same decade a long breakwater, called Dog Bar, was built of granite westward into the mouth of the harbor with a navigation light at its extremity.

In the middle distance toward the center city is Ten Pound Island Light on the eastern side of the harbor. So the lights came first.

So far as life-saving was concerned in the earlier days there were Massachusetts Humane Society huts of refuge and later boathouses with lifeboats manned by local fishermen volunteers.

From the beginning of the United States Life-Saving Service the first station in that area was the one on Plum Island some miles south of the Merrimack River. That was followed by the one at Davis Neck, Annisquam.

The twin lighthouses on Thatcher's Island were built in the 1860's. Each is 124 feet high and built of stone masonry. In 1932 during the Great Depression the northern light was turned off. Before ten o'clock at night, November 26, 1898, the keeper saw the steamer *Portland*, heading north for her rendezvous with history.

When the service closed the Davis Neck Station, it built a new station at Land's End, Rockport and named it the Straitsmouth Life-Saving Station. Its area of service guarded the stretch of water from Plum Island to Gloucester.

The last U.S.L.S.S. station in the Cape Ann area was built at Dolliver's Neck on the west side of Gloucester Harbor.

The Life-Saving Service established this Straitsmouth Station in Rockport and then closed the one at Davis Neck, Annisquam.

Nine months before the *Portland* Hurricane a severe northeast storm struck Cape Ann and drove the steamer *Frank Scripture* onto these rocks at Pigeon Cove. The loss was total: $20,000.

Into the early 1880's along the Massachusetts coast there were life-saving stations at Plum Island and Annisquam with the third station at Fourth Cliff, Scituate. From Gloucester south past Boston Harbor and Point Allerton at Hull there were no federal government life-saving stations until the south end of Fourth Cliff in Scituate. By 1891, however, there were added stations at Point Allerton, North Scituate and two more straddling Plymouth, Gurnet to the north and Manomet Point southeast.

In the open Atlantic off Rockport many vessels have foundered in the severe weather of northeast storms. One Saturday morning in September, 1854 a lumber schooner, the *C. H. Sampson*, from Georgetown, South Carolina, sprang a leak and filled with water three miles east of Thatcher's Island. The surge of the ocean shifted the deck load and started the planks. Lucky for the crew there was another schooner near by, the *Emma Prindle*. She closed quickly and took everyone off the stricken vessel and carried them to Gloucester. The *Sampson* sank so fast that no one on board had time to retrieve any of his possessions.

Pushed by gale winds the *Sampson* drifted ashore near Bass Rocks, Eastern Point, where she remained stranded, thumping very hard and eventually going to pieces. Her cargo of 80,000 feet of hard pine was scattered widely and most of it lost.

The schooner *Sampson* was only three years old, 140 tons and well found. She was valued at $6,000. and insured for two-thirds of that.

Humane Society volunteers could never have reached the stricken vessel from Pigeon Cove in that wild weather, so lifesaving took place from a passing schooner, the *Prindle.*

The mountainous waves drove this schooner ashore at Pigeon Cove to her destruction. The wreck later sold for just seventy-five dollars.

The storm that lashed the Massachusetts coast that is recalled the most is the *Portland* Storm of Thanksgiving weekend, Saturday through Monday, November 26-28, 1898. The incredible violence of those two cyclones that combined off the coast into one surge of fury drove ten vessels ashore or sank them at Rockport and Pigeon Cove. The Pigeon Cove losses were the heaviest with seven ships wrecked on the rocks or sunk at their piers. Most of those schooners were there to load and move granite from the local

Two schooners, a steamer and a barge are on the rocks and underwater in this *Portland* Hurricane aftermath at Pigeon Cove Harbor. This loss was $21,500.

quarries, the major industry. The other three vessels were destroyed at anchor in tiny Rockport Harbor.

Four years almost to the day a British steamer, the *Wilster*, stranded on a beach three miles south of the station during a southeast gale with thick weather and a high sea. The keeper sent a surfman to hire a horse to pull the beach apparatus to a point opposite the steamer. The surfman finally returned with a horse that he had had difficulty acquiring, as many people did not want their animal out in such weather.

Upon arrival at the shore the apparatus was set up and a shot fired from the Lyle gun which landed perfectly among the crew amidships. As the steamer was rolling and pounding her decks were swept with waves rolling over her, the crew had great difficulty making fast the tail block and hawser.

When everything was ready, the work of rescue began, but after two men were landed the operation had to be temporarily suspended, because it was impossible to keep the hawser taut, as the sea and the incoming tide were pushing the steamer farther onto the beach. When she was lying more quietly, the work of rescue began again. The remaining men on board-twenty-one of them-were landed one at a time by the breeches buoy. They were sheltered nearby at a private home where they also dried their clothes. The next morning the sailors from the steamer were taken to the life-saving station where sixteen of them received shelter for three days.

After the gale subsided, the officers and crew of the steamer went back aboard to try to get her off the beach. Nine days later on March 9, 1902 a signal was displayed on the steamer and the surfmen and keeper returned to the beach where they were asked to take the crew and some wreckers off the vessel.

The surfmen put a line on the steamer, using a heaving stick, and twenty men were landed on the beach in the ship's boat, the surfmen hauling the boat back and forth from the shore to the ship.

When the weather permitted, the wreckers continued to work on the vessel, removing the cargo to lighters. Two weeks later the steamer floated and tugs hauled her to Gloucester for temporary repairs.

After the steamer *Wilster* grounded three miles south of the Straitsmouth Station, surfmen took the crew of twenty-three off by breeches buoy. The steamer was floated later.

Three of the crewmen needed clothing and they received it from the stores of the Woman's National Relief Association. Railroad tickets were obtained for free transport to Boston for sixteen men. After the gale that forced the *Wilster* to strand had subsided, the weather never again during that time was so severe or she would have been pounded to pieces on the beach. The lifesavers kept one surfman on the scene the whole time, twenty-three days, to report quickly to the station in case of serious trouble.

The Straitsmouth Station had as many minor rescues as those elsewhere along the Massachusetts coast. Two calls for help in 1902 off Rockport are typical. In late July the American schooner *Metropolis* stranded on the bar while attempting to pass between Milk Island and the shore. The station keeper who was alone on duty during the off season boarded the vessel and assisted to keep her afloat and into deep water, whereupon the schooner proceeded to her destination, apparently undamaged.

Despite years of attempts to build a nine-thousand foot long breakwater to make Rockport a harbor of refuge, it was not until 1884 that work began. This is the first one-hundred feet.

Bottom: Eventually some six-thousand feet of foundation stone was placed and it was capped by six-hundred feet of granite blocks. These federal inspectors are looking over the plans and the work.

In late November the American schooner *Newell B. Hawes* which had survived a number of misadventures around Plum Island, sprang a leak while beating around

Thatcher's Island, anchored in Sandy Bay, a half mile northwest of the station. She set a distress signal, which was answered by surfmen who pulled to Rockport in search of a tug. Finding none, they returned to the station and telephoned to Gloucester for one. Later in the day the surfmen boarded again in response to a signal, manned the pumps, got vessel under way, and started for Gloucester, but the high seas compelled them to seek shelter nearby in Pigeon Cove. Once there the schooner was moored to a wharf and everything made secure. The *Newell B. Hawes* had to have the many lives of a cat, as she escaped disaster again.

Around Halloween the following year the American schooner, *Ethel F. Merriam*, took a heavy squall while working into Sandy Bay and dropped her anchor but then dragged aground at Gap Head northeast of the station. The surfmen boarded the vessel and made sail, then dragged the anchor into deep water, where the schooner lay in apparent security; but after noon the wind and sea increased, and, in response to a signal of distress made by the master, the surfmen procured a tug, which towed them in a surfboat to the schooner. The surfmen then boarded the vessel, hove up her anchor, and remained on board until the tug towed her into Rockport Harbor.

Within a few weeks the American schooner, *Laughing Waters*, was off Thatcher's Island when her foremast was carried away and she entered Sandy Bay and anchored. At the request of the master, the keeper telephoned to Gloucester for a tug, which came and towed the schooner out of the harbor to Gloucester for repairs.

A limer from Maine grounded on Londoner Ledge one mile southeast of the station and a thousand feet off Thatcher's Island the following year in September. The surfmen from Straitsmouth saw the peril to the schooner, *Hastings*, and went to her where they took in her sail and threw overboard her deckload of lime casks to lighten her. Then they went ashore and called a tug. They returned to the *Hastings* and found her leaking and the lime below decks smoldering. Mortar was mixed and all hatches and other openings plastered shut with the view of smothering any fire that might start, the usual fate of limers. The surfmen turned to the pumps and stayed by the schooner until a tug came and pulled her free. The Londoner Ledge is a half-submerged reef, a dangerous obstruction.

Early on a February morning in 1906 two men in a dory under sail for Thatcher's Island were blown to sea by a heavy offshore wind, and in danger of being lost. The keeper, seeing their plight, launched the surfboat, pulled to their assistance, and towed them safely to shore. That time of the year is dangerous for such escapades off shore, especially in a heavy wind.

Within a month a tragedy occurred off shore at Rockport in the sinking of a tug from Portland, Maine with the loss of three of her crew. The vessel was the steam-powered *Willard*, only 50 feet long with a beam of 13 feet. The danger to the crew was in her freeboard which was a mere 14 inches above the water amidships. The *Willard* was bound from Boston to York, Maine, carrying on her deck a four-ton bucket for a dredge and a large coil of five-inch diameter hawser. Part of the dredge bucket was on the front deck and the other part along with the coil of hawser was stowed on the after deck.

The hills back of Rockport and Pigeon Cove were a mass of derricks, guy wires, steam hoisting engines and hundreds of workers bringing stone out of huge holes in the ground. After that, railroad locomotives hauled it on flat cars to the water.

The *Willard* left Boston on the afternoon of February 27, 1906 and made slow progress along the coast against strong northerly winds and threatening weather. Her captain, E. R. Sawyer, decided to seek shelter at Gloucester Harbor. The weather did not improve until the morning of March 1, so the voyage continued at 5 a.m.

So far the trip was uneventful until she rounded Straitsmouth Island and headed for the inside of Avery Ledge where she ran into a strong head wind and a considerable sea. Then the vessel started to ship spray.

On the port quarter lay Sandy Bay with Rockport Harbor just two miles away. Despite the wind and the sea Sawyer decided to keep going on his course and head for Halibut Point, at the northerly end of Cape Ann. As the *Willard* reached farther out into the open water the wind, which was now on the port bow from the northwest, struck the tug with increasing violence and the seas more formidable. By this time the captain decided to make the harbor at Halibut Point.

Without warning a heavy gust of wind struck the little tug and pushed her to starboard, submerging the rail and bringing the water to the fireroom doors. To bring the vessel back head to the wind the captain put the helm hard astarboard and she fell off with the wind and sea full on her starboard side. The waves reached the tip of the deckhouse doors, flooding the fireroom and the engine room. Then nothing could be done to save her from sinking. The crew of five was in various parts of the vessel: the captain and the deck hand were in the pilot house, the cook was in the galley underneath

Once down on the Rockport Granite Company's pier the stone was lifted into the hold of vessels like the big, Boston schooner, *Mary H. Diebold,* and taken to Miami and Key West for construction projects.

the pilot house in the forward part of the deck house, and the firemen and the engineer were below decks in the fireroom and engine room.

The captain sensing the danger ran out of the pilot house with the deck hand and shouted a warning to the others. Just back of the stack was a dory, the only small boat carried. The firemen and the engineer came out of below decks where they were working and reached the dory just ahead of the captain and the deck hand. The captain cut the lashings on the dory and it swung in the wind and fell to the water right side up. At once the fireman leaped into the dory, and the captain followed only to fall into the water from where he quickly grabbed the dory and climbed in as the wind and sea swept the dory away from the sinking vessel.

The two in the dory cleared the oars and started pulling for the sinking tug, but the wind was too strong and they kept losing ground. They turned and headed for the shore and pushed by the wind they landed some one-thousand feet from the Straitsmouth life-saving station.

Surfman Griffin, standing the morning watch in the lookout tower, was the first person to sight the *Willard* as she rounded Thatcher's Island at 6:20 that morning. Griffin left the tower for breakfast after being relieved by surfman Stanwood. Just before 7 o'clock Griffin returned to the tower. Stanwood later testified that he first saw the *Willard* when she was steaming around Straitsmouth Island. When Griffin returned he testified he looked for the tug and she was no where to be seen. He assumed she made harbor at Rockport.

Up to this time the keeper did not know the *Willard* had been sighted; he was concerned about a barge anchored off the salvages. The only other object on the water was a small boat with two men in it. That looked like real trouble to the keeper, as in his experience an open boat so far out at such an early hour portended disaster to a larger vessel. While the keeper and surfman Griffin were watching, a telephone call from Rockport informed them that a tug had capsized inside the Sandy Bay breakwater.

The keeper sent four of his men in the station's dory, which could be launched faster than the surfboat. The keeper set out alone along the beach to assist the two in the small boat approaching. The surfmen in the dory circled the area off shore where the *Willard* sank while the keeper helped the two survivors from the *Willard* out of their dory and back to the station for shelter and warmth. Captain Sawyer was in serious condition with his hands, knees and feet frost-bitten. The keeper treated Sawyer and helped him recover. The fireman was in better shape as he hadn't fallen into the water, as Sawyer had. The air temperature was 7 degrees above zero. The condition of the sea coupled with the strong northwest wind blowing directly on shore would have prevented the surfmen from reaching the scene in less than a half hour, and it was doubtful whether those who drowned could have survived that long in the frigid water. Nothing was seen in those frantic moments when the tug lurched of the cook who shut himself in the galley to keep out the water.

The engineer, the deck hand and the cook died in the sinking. At an official hearing into the tragedy, it became clear that the *Willard* sank in the few moments that the change of watch in the tower took place. Stanwood admitted he didn't keep an eye on the tug the whole time. He testified that the *Willard* was there the last time he looked. Stanwood's failure to exercise constant vigilance earned him a severe reprimand and a thiry-day furlough without pay. His punishment would have been dismissal, except for his previous good record, long service and other circumstances in his favor.

Almost two years after the steamer *Wilster* stranded, another steamship the *Dudley Pray* stranded on the south end of the Sandy Bay breakwater a mile and a half northeast of Straitsmouth Station. The lifesavers pulled to the scene in a surfboat and went on board after which they went ashore and wired the owners who telephoned for tugs. When the tugs arrived the surfmen helped throw coal over the side and ballast as well as running lines from the steamer to the tugs. They stayed aboard while the tugs hauled away at flood tide. Despite the power of the tug engines, the *Dudley Pray* stayed aground and would not move. Her hull was badly damaged and had filled with water. The surfmen transferred the crew of eleven to one of the tugs which carried them into Rockport Harbor as the owners decided to abandon the steamer where she lay. A wrecking company eventually removed the engine and boilers and anything else of value. The hull was badly damaged and so the wreckers blew it up with explosives.

Later that year in October a gasoline launch with two fishermen from Maine capsized in heavy seas off Rockport, drowning the two occupants. Nothing was known of the case since there were no witnesses to the nighttime tragedy. The two were headed from Cape Porpoise, Maine just above Kennebunkport to Boston with a large cargo of lobsters. A fresh wind had kicked up heavy seas during the night when the accident happened. Local fishermen found the small launch sunk off the Sandy Bay breakwater with just her stern above the water. It appeared that she was trying to make harbor at Rockport because of the rough weather.

The launch was not suited for open sea voyaging under any circumstances. She had no decks or cover except a small house over the gasoline engine. In addition she had a heavy load of lobsters which made the open ocean trip even more hazardous. The keeper at the Straitsmouth station believed the launch either struck a submerged ledge or the breakwater and then foundered. The keeper checked the details of the night patrol and found it to have been well covered. No outcries were heard or signals seen that would indicate that a vessel was in distress. At the official inquiry afterward it was brought out that many of the fishing vessels sailing off that coast did not carry lights or display any, by itself a dangerous practice. The launch in question sank a mile from the station-too far away for any one in the lookout tower to see unless she had displayed signals.

The active season had just begun in August when a U.S. Navy tug, the *Nezinscot*, sailing from the navy base at Portsmouth, New Hampshire to Boston, capsized and sank seven miles from Cape Ann and off Halibut Point to the north of Ipswich Bay. Two members of her crew were in their berths at the time and they drowned. Seven others got into her boat and they rowed for Lanesville where they landed at the cove. The other four clung to the wreckage in the open sea, but two of them were unable to hold on long enough and were drowned.

When the sailors reached Lanesville they telephoned the Life-Saving Station at Straitsmouth which also telephoned Gloucester where there was a power lifeboat which raced to the scene, arriving in time to rescue the last two survivors. The lifesavers applied first aid to the two and they were landed at Lanesville and placed in the care of a doctor.

In the meantime as soon as the call came from the seven survivors at Lanesville the surfmen and keepers at Straitsmouth launched their boat and pulled for the scene. They arrived after the power lifeboat rescued the final two. By 1910 most stations were issued power lifeboats.

Some coastal tugs were designed for open ocean work, but others had a low freeboard amidships and were not as safe in a heavy sea. The *Willard* and *Nezinscot* were like that.

Sailing down the coast this night the captain of the schooner and his mate try to figure out where they are on a chart. It was called dead reckoning. The captain's wife shows her concern with folded hands and tight lips. She has probably made many such voyages.

The schooner *Winnie Lowry* was heading south past Rockport with Maine lumber when she was driven onto these rocks and destroyed. Her crew was rescued by surfboat.

Here is a frightening transfer in a wild sea.

Beach patrol had its hazards everywhere, but none worse than a rocky shore. The lantern helped surfmen to survive, especially in winter with frozen spray as footing.

As elsewhere the Straitsmouth surfmen rolled out one of their surfboats and posed with it on a pleasant day. The other two and the keeper are missing. A summer hotel, the Straitsmouth Inn, is in back.

GLOUCESTER

America's most famous fishing port is located on Cape Ann some 36-miles north of Boston. Men from Gloucester sailed out to the Atlantic to fish, beginning in 1623. A great series of emergencies for Gloucestermen and other seafarers along the 192-miles of Massachusetts coastline took place in December 1839 when three storms struck in succession that month.

On the 14th a full gale followed by another eight days later. Five days later the area was lashed by a devastating hurricane.

In Gloucester alone, fifty vessels were lost in that first storm. It is difficult to get an accurate count of vessels that simply disappear at sea, known as "gone missing." Historian J.W.Dalton said the toll for the three storms was one bark, four sloops, 17 brigs, 68 schooners and more than 200 lives lost.

Besides vessels totally wrecked, there were five sloops, 22 brigs, 23 ships and 168 schooners driven ashore or crippled in other ways during those titanic storms.

The storm of late November 1898, raged from Saturday, the 26th in the evening through the weekend until it finally moderated on Monday late in the day on the 28th. The worst hit was southern New England with the Massachusetts coast the hardest hit with its worst impact from late Saturday through Sunday night. It was one of the greatest storms ever to strike New England. Afterwards it came to be called the Great Gale of 1898 and sometimes the *Portland* Gale. The word gale was used with the other words. Yet there were wind measurements indicating a velocity over gale intensity and of violent hurricane speeds. Yet through studying the newspapers and other publications there is almost never encountered the word hurricane. But the tempest was surely a most violent hurricane.

Just inside the western edge of Gloucester Harbor was the U. S. Life-Saving Station at Dollivers Neck, south of Stage Fort Park. The staff on the left carried signal flags for ships. The U. S. Signal Service flew storm warnings there.

At Gloucester there were 28 vessels destroyed. Across Cape Ann a few miles to the north at Pigeon Cove seven more were lost and at Rockport three more. At Salem four vessels were destroyed and one at Manchester. The toll for the North Shore of Massachusetts was forty-four lost vessels.

Surfmen land survivors of a shipwreck on a Massachusetts beach.

In late November, 1902 a small schooner carrying cement from Boston to Portland, Maine stranded on the rocky shore of Eastern Point two miles southeast of the Gloucester station during a high sea. The keeper at Eastern Point Light saw the vessel, *Beta,* go ashore and he raised a flag to signal the lifesavers of the accident.

The surfmen pulled to the scene and found the sea so rough that it was not possible to reach the wreck by rowing there. So lines were rigged to improvise a breeches buoy which, after several attempts, was put on board the schooner, the wife and five-year old son of the captain were safely landed.

The schooner was breaking up so rapidly there was not enough time to rescue the crew the same way. The crew was saved by having them jump into the sea and were safely hauled ashore by lines thrown to them by the surfmen. A few minutes after the last sailor was rescued the *Beta* broke in two and went to pieces. The captain of the schooner, his wife and young son and a crew member spent the night at Eastern Point Light. The other three sailors were returned to the station in the surfboat and were given dry clothing from the supply donated by the Women's National Relief Association.

The stranding happened after 8 o'clock at night and the next morning the surfmen rowed over to Eastern Point Light and brought back to the station the four people from the schooner who were there. They, in turn, were given dry clothing. Two of the rescued were given shelter for one day and the other five, including the captain and his family stayed at the Gloucester station for five days.

The loss of vessel and cargo was total: $1,200 for the schooner and $1,000 for the cargo.

Two weeks later and not far from the wreck of the *Beta* the British schooner, *Flo F. Mader* stranded on Dog Bar Breakwater after 2 o'clock in the morning in a snowstorm.

A surfman on patrol discovered her and burned a Coston signal to let the men on board know that they had been seen and help was on its way.

The surfmen pulled to the vessel where the captain asked that they go ashore and get a tug to come out to haul them off. The tug arrived and the surfmen ran a hawser to the schooner. The line parted and the operation had to be put off for a more favorable tide.

As the wind and the sea were increasing, the surfmen took the personal effects of the master and crew to the station for security, and also three of the crew who feared to stay by their vessel.

The rest of the crew abandoned her later, leaving the master and mate on board. The surfmen kept constant watch upon the stranded schooner until the next afternoon, when two tugs came out from Gloucester Harbor and got her afloat. The lifesavers boarded the schooner and performed necessary services while she was being towed to a marine railway in East Gloucester. The lifesavers returned to the station towed by a tug.

Later they carried back to the schooner the personal belongings of the crew, all of whom returned on board.

A year later in September the American schooner, *William M. Walker*, stranded on Dog Bar Breakwater, trying to enter Gloucester Harbor. The keeper and crew went to her and found she was pounding heavily and leaking. They took the personal effects of the crew to the station, and beached her in a safe place.

The surfmen took the crew back to the station and gave them dry clothing and shelter and fed them overnight. The next afternoon the damage to the schooner's hull was repaired and the crew and their belongings returned to the vessel and sailed on the next tide.

A schooner at anchor in Gloucester Harbor on a January morning was water-logged and abandoned. The lifesavers called a tug company and met that vessel at the scene. The tug hauled the schooner, *N. Jones*, to a Gloucester dock. Later her crew of four men came to the life-saving station where they were sheltered for two days.

On an early August morning in 1905 the American schooner *Moses Adams* stranded on Whale Back Rock while getting under way in a fog three and a half miles east of the life-saving station. The surfmen went to her and ran out anchors, and with buckets kept the water rising in her hold from reaching the fires, her pumps being unequal to the task. She was finally floated, and with the 15 people on board towed to Gloucester by a tug with the surfmen following in case of more trouble.

On the last day in February of 1906 the sloop,*Cora,* stranded on Eastern Point near the breakwater. It was the middle of the afternoon during a fresh northeast gale and the vessel had parted her moorings.

An old sailing vessel, a victim of New England weather, lies on the Massachusetts shore pounded by waves that for years rise at every tide through her broken side.

A breeches buoy rescue with the line sagging to expose the sailor to the top of the surf. A surfman wades in to help the man reach shore and climb out safely.

This crowd is opposite a wreck and just sending back an empty breeches buoy for another survivor. Some of those rescued are standing in the back next to that surfboat. Here the tension remains high until all those in danger are safely on this beach.

A young girl surrounded by the arms of a rescuer has a breeches buoy ride over high seas to safety, accompanied by her suitcase. She will always have an exciting story to tell.

In nice weather on Sundays visitors showed up at many stations. In the active season families of the crew spent much of the day with a picnic. Young women appeared and were given rides on a breeches buoy by the drill pole, left. They were always welcome for this aerial roller-coaster, as the single surfmen were lonesome and the girls received a great deal of attention and offers to come again.

The keeper launched the Race Point surfboat and pulled to the scene. They found the sloop wedged among the rocks. The surfmen removed her ballast, lashed empty casks under her bilge, ran out her anchors astern and a line to the breakwater. Some of the outlying rocks were removed, and at high water she was pulled afloat and taken to a safe anchorage by a tug.

The crew of the sloop was exhausted, so the keeper took them to the station where they received dry clothing and meals and an overnight stay. The next day the surfmen assisted to replace the ballast and the *Cora* sailed to her destination.

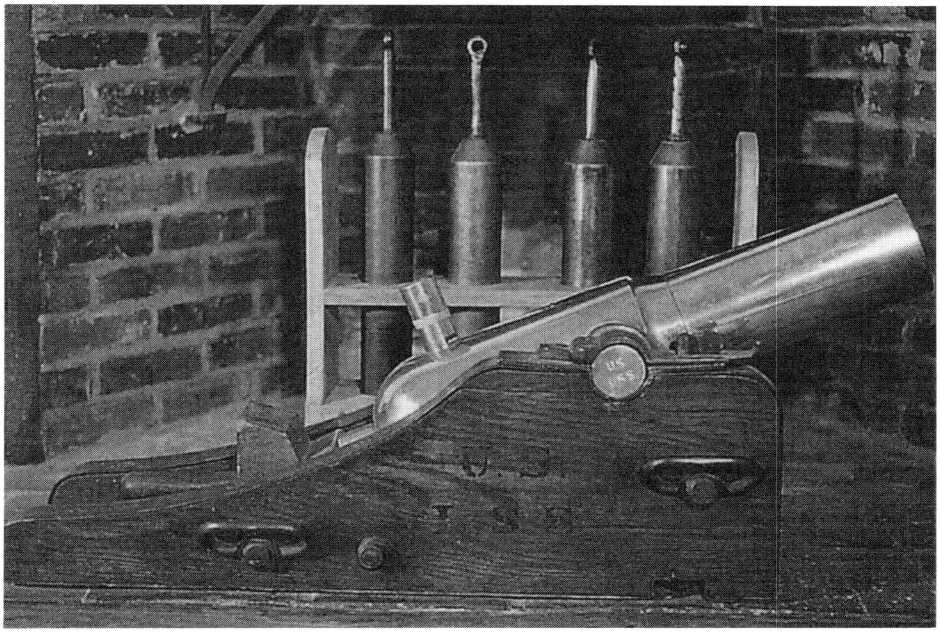

David Lyle's brass canon developed at the Springfield Armory became standard issue at the U. S. Life-Saving Service and later the Coast Guard. The gun's projectiles are in the back. Although the range of these cannon was higher, the breeches buoy was used mostly up to six-hundred yards. Lifesaving historian, Richard Boonisar, owns this gem.

Some of the worst weather recorded along the Massachusetts coast has occurred in April. If April is the cruelest month as referred to by poets, then certainly accidents to coastal shipping had to be part of the reasoning.

An April wreck of a vessel at Gloucester happened during a southeast gale in thick, rainy weather on April 10, 1906. The American schooner, *D.Gifford* started to drag while riding at anchor and was driven by the gale onto Field Rocks about one mile north-northeast of the station. She was discovered there by the surfmen on patrol. At the station a signal was set for a tug, and the keeper telephoned the tugboat company in Gloucester. Then the keeper launched a dory and with his crew rowed to the schooner to assist, as she was lying stranded in a bad place.

Two tugs were sent, but after rounding Dog Bar Breakwater, they turned back as the sea was running too high to continue safely.

The surfmen, however, kept on and finally reached the stricken vessel and went aboard. By that time she was rapidly filling with water and her skipper decided to abandon her. The surfmen stowed the effects of the sailors in one of the ship's boats which they launched. All six of the crew got into the dory which then towed the ship's boat away from the schooner. While fairly nearby to the schooner *Gifford* but heading back for the station, the lifesavers and the rescued sailors turned to see the ill-fated craft give a deep lurch to starboard and settle on the bottom in shoal water.

Around Thanksgiving of 1907 the British schooner, *Cora B.*, parted her anchor chains in a northeast gale and stranded at 2:30 in the morning one-half mile north of the station in Gloucester Harbor. The lookout saw the vessel and alerted the crew and the surfboat was launched and pulled to the scene.

When the surfmen reached the schooner they found her abandoned by her crew in their own boats. The sailors were found on a nearby beach, wet and cold, and were taken to the life-saving station where the five of them stayed for two days after which the lifesavers took them to Gloucester in the surfboat.

On the second day after the stranding the lifesavers went on board and recovered the crew's clothing. Then the surfmen and the ship's crew saved part of the vessel's cargo and brought it back to the station and stored it in boathouses.

During a thick snowstorm in February, 1908, the keeper received a request by telephone from the Boston and Gloucester Steamboat Company who felt anxiety since their vessel was late on her regular trip from the Hub. When she finally cleared Dog Bar Breakwater and could be sighted from the lookout, the keeper telephoned the company office in Gloucester with the news.

During April in 1908 the surfmen of Gloucester responded twice to emergencies. On the 15th a gas launch dragged her anchors and stranded on Long Beach one mile northeast of the station. The craft was discovered by a man on beach patrol after midnight. The surfmen went to the scene in a small boat and ran out an anchor, but the tide was too low to get her off. They returned at high tide and got the launch afloat undamaged. On the 16th the Gloucester surfmen responded to a call from the Plum Island station to help with a stranded schooner, the *Gatherer*, in Essex lower harbor off Ipswich Bay. The Gloucester crew sailed around Cape Ann in their power lifeboat, while the Plum Island lifesavers rowed down in their surfboat.

Upon arrival they found the schooner high and dry, badly strained and leaking. The lifesavers ran out an anchor, lightered part of the cargo, and on the rising tide kept her pumped out. At high water the surfmen got her afloat and took her to a safe berth farther up in the harbor at Essex.

On a March afternoon in 1909 the Gloucester lifesavers had two emergencies simultaneously. On their way in the power lifeboat to a disabled gas launch, they discovered a schooner anchored in a dangerous position about two miles southeast of the station. One surfman was placed on board the launch to repair the motor, while the lifeboat proceeded to the schooner, the *Eben Parsons*. A line was put aboard the schooner, the anchor was raised and the surfboat was used as a tug which then towed the vessel to a safe anchorage.

The Gloucester lifesavers had a three-day odyssey to find an auxiliary schooner, the *Reliance*, which struck on southeast breakers at 6:30 p.m. on October 1st. Her crew abandoned ship six miles southwest of the station and proceeded to Gloucester. Of the eight member crew two of them came to the station and reported the wreck. They were taken in the power lifeboat to the spot where the vessel struck, but she had drifted away and vanished.

The next morning the power lifeboat put out with three of the schooner's crew and did locate the schooner sunk on Salt Rocks off Manchester. However, nothing could be done in the way of salvage, because the tide was too high.

At 4 o'clock that afternoon the surfmen returned with some crew members and took off all her sails.

The next day at 4:30 a.m. with the tide right the surfmen returned to the scene for the fourth time and helped a wrecking company to raise the schooner and bring her to Gloucester. On the way the wrecking pump broke down and the lifesaving crew kept her hull dry by bailing her out with buckets while she was being towed into Gloucester.

The stone schooner *Multnomah* stranded on Bass Rocks, East Gloucester, in June, 1910 on a night voyage for Lanesvillle. the crew of five were taken off by surfmen from Gloucester and one of them spent a day at the station. It was a clear night with no problem from the weather. William Robb Cheves, of Lanesville, the vessel's owner, thought the crew had visited a Gloucester tavern earlier. The *Multnomah* was a total loss.

The lifesavers at Gloucester had a number of minor emergencies in 1910 and two major ones. The sloop, *Chicnook,* struck the end of Dog Bar Breakwater and sank in 20 feet of water. It was May 30th and the crew of three escaped to the breakwater where they were rescued by the surfmen in a dory. The station crew in the power lifeboat returned to hook a grapnel under the bow and by careful operations raised the sloop and made a hawser fast around the end, and with the help of two private launches towed her to East Gloucester where she was tied to a wharf.

On June 16th a granite schooner from Lanesville, on Ipswich Bay, stranded on Bass Rocks on a clear night. She could not be salvaged and was a total loss. She was, the *Multnomah* on a trip from Gloucester around Cape Ann to Lanesville where she regularly loaded granite for shipment by her owner, quarryman William Robb Cheves of Lanesville. The *Multnomah* grounded on rocks three and one-half miles northeast by east of the Gloucester station. She was traveling light with a crew of four, one of whom was returned to the station where he stayed for shelter one day. The two-masted schooner was valued at $3,500 of which $1,900 was estimated to be the loss. The vessel was not insured. The grounding and destruction of the *Multnomah* was never successfully explained as the night was a mild one with no pressure from the elements. Too much alcohol consumed at a Gloucester tavern was one explanation.

A collision in Gloucester Harbor on March 14, 1911 killed the entire crew on a fishing launch about two miles from the Gloucester station. The launch was the eleven-ton, *Hope*, and the other vessel was the schooner *Hattie A. Heckman*.

The *Heckman* lowered boats immediately and picked up two of the four victims and took them on board. The two were apparently dead when recovered from the water, but whether or not the crew of the schooner made any effort to revive them is not known. The Gloucester lifesaving crew boarded the schooner and took the bodies in hand, working over them for an hour assisted by three doctors summoned from the shore by signals. However, the two could not be revived.

The collision was witnessed by a surfman in the station watch house who testified at a hearing, which the service always conducted after an accident involving death: "I saw the schooner, *Hattie A. Heckman*, being towed out of the harbor. After the tug dropped her she tacked toward the station. The launch, *Hope*, came around the breakwater, standing in. Then I looked toward another boat that had attracted my attention. When I looked back the *Hope* was gone and a lot of wreckage was hanging on the bobstay of the *Heckman."*

The surfman, Haley by name, gave the alarm and the station crew left in the power lifeboat. Besides trying to revive the two victims the keeper, Nelson F. King, questioned those on board to find out how the tragedy occurred. They brought the bodies of the two men ashore, but did not learn until they got back that there were two more victims unaccounted for.

The lifesavers located the launch in 35 feet of water the following day and they placed a buoy to mark her. Five days later a salvage firm raised the *Hope* and towed her to a dock in East Gloucester. They found the body of the third victim.

Two and one half months later a body identified as the remains of the last victim was discovered near the end of Niles Beach about a mile and a half east of the life-saving station.

Two and a half years after that accident a similar one happened in Gloucester Harbor on September 28,1913. Again it was a collision between a launch and a schooner. The scene was less than one mile east-northeast of the Gloucester Life-Saving Station.

The launch, *Abbie E.*, was one of a fleet of boats that visited Gloucester on Saturday afternoons. The five members of the party aboard set out for Boston after midnight.

The launch had gone only a short distance when the collision occurred with the schooner *Rhodora.* The man in charge of the launch said he saw the lights of the schooner, but that he misjudged the distance between the two vessels in the darkness. The launch crossed the schooner's bow and the larger vessel struck the other amidships, severely damaging her below the water line and carrying away most of her cabin, where the three men who died were at the time. The two who survived were on deck and were rescued by men on the schooner. The *Rhodora* sailed into Gloucester Harbor and the two survivors were sheltered on board until morning. The local police were notified of the accident and they called the Life-Saving Station.

The keeper took his crew out in the power lifeboat with a dinghy in tow and finally after a long search located the sunken launch in 33 feet of water.

The lifesavers buoyed the wreck and went to the Gloucester wreckmaster, who employed a diver to look for the missing bodies. The diver could find none. Some time later the body of one of the men was found when the launch was raised. The other two were never found.

The Gloucester Station at Dolliver's Neck flew a brand new United States Coast Guard flag in 1915 after the United States Life-Saving Service was absorbed into the Coast Guard. The hundreds of other stations also got new flags. However, it was several years before the keepers and the surfmen had new uniforms. That transition took time. The work of lifesaving on a dangerous coast remained a hazardous occupation.

A northeast storm drove the schooner *Mexican* onto the rocky shore of Cape Ann near Lanesville. The vessel was carrying coal from New York to Bucksport, Maine when she wrecked in October, 1890. Lifesavers from the Davis Neck Station and from Gloucester took the crew of four off and to safety. Most of the cargo was saved, but not the ship which broke up.

MAINE: THE SHIPS AND THE CARGOES

Maine was a big maritime state. In the town of Rockport from 1844 to 1892 Carlton & Norwood, shipbuilders, built 63 ships for their own account-schooners, barks, brigs, barkentines, full-rigged ships, an entire fleet, save one, built by John Pascal, a master Maine shipbuilder.

Camden Anchor Works was one of eight forgers of large ship anchors in the United States. Camden produced more than the others combined. Other anchor works were at Bath and Calais. In Camden oakum was made from old cordage. Windlasses, capstans, blocks, treenail fastenings and other items were supplied by the Knowlton Bros. Foundry. Rigging and sails were made by E. S. Bohndell of Rockport; the duck sailcloth used for some sails was woven at Westbrook.

A major ship facility was a dry dock at Knightville, Cape Elizabeth (now called South Portland). There were two basins in that dry dock.

A huge fleet managed by Captain Jacob S. Winslow had eighty-five sailing vessels in the 1870s-80s, mostly schooners and barks.

Titcomb & Thompson were shipbuilders who, despite financial depression, built in 1873 and '74 seven three-masted schooners, one barkentine, and four full-rigged ships. All told, Captain Thompson, his full name was Nathaniel Lord Thompson, was instrumental in building more than one hundred vessels, some of them at Kennebunkport.

In 1874, 21 full-rigged ships, 51 barks, seven barkentines, 61 brigs, 135 schooners, 37 sloops-including big stone sloops-27 steamers and 81 vessels under 20 tons had Portland as their home port.

So much of the lumber arrived by water. Treenails made of North Carolina locust were used in sailing vessels' hulls. Planking and ceilings on schooners were southern hard pine. Frames were of Virginia white oak. Lower masts and yards were of Oregon pine.

On sailing vessels built around Camden the capstans, pumps, windlasses were products of Knowlton Bros. Foundry and Machine Works in Camden.

Down the coast The Portland Company was founded in 1846 by local people to build steam locomotives for the projected Atlantic & St. Lawrence Railroad. However,the company did a big business in ship propellers and marine engine forgings. With all these allied industries, Maine was one of the greatest centers of shipbuilding in the western world. Maine had hundreds of miles of coastline with rivers and bays and inlets, vast forests with a variety of timber and a hearty group of men dedicated to the work ethic. Shipyards that built wooden vessels didn't need much: a plot of land that was firm and could stand extra weight, located right on the shore with fairly deep water and a supply of timber. Bath, on the Kennebec River, became the most famous and, in fact, is still building ships, now almost exclusively for the United States Navy. Most of the many other yards closed years ago. Bath built its first ship in 1783, a schooner of 99 tons. Other places had been building ships for many years before that. But Bath became

Bath Maine was "The City of Ships." These workmen paused for a photograph in the Kelley, Spear yard. Behind them is a schooner-barge under construction.

This is the yard of the Dunn and Elliot Company at Thomaston. These men are building up the keelson in the schooner *Republic* which will be turned over to its owner, a New Bedford dealer in whale oil.

The *Bertha L. Downs* is ready for her launching into the Kennebec River. Some of the guests are aboard and others are gathered near her stern. The donkey engine has steam up to hoist sails. The *Downs* was built for a New Haven, Connecticut firm.

the King Kong of New England wooden shipbuilding and gained the title, "The City of Ships."

There were yards in many other places. Also on the Kennebec were yards at Woolwich, Bowdoinham, Dresden, Richmond, Gardiner, Hallowell, Augusta, Vassalboro and Phippsburg. Other yards were at Kennebunkport, Boothbay, Harpswell, Georgetown, Pittston, Brunswick, Perkins, Rockland, Camden, Belfast, Rockport, Bangor, Searsport, Stockton Springs, Thomaston and Cape Elizabeth. I may have missed a few-maybe more than a few.

There were others, too, they seem to have been in almost every town on the water. Their production was huge and sailed all over the world, those Maine-built ships.

Most master shipbuilders at Bath owned shares in the vessels they built.

The cost of building ships at Bath was kept relatively low by the system of operation: low wages and the efficiency of the men.

The principal jobs were contracted to various master workmen who then hired other workers and moved them from yard to yard.

The carpenters, shipwrights, joiners and caulkers supplied their own tools; the contractors-fasteners, blacksmiths, riggers and sailmakers provided their own outfits. So the average yard owner did not maintain a large labor force or a stock of tools.

In an 1882 U.S. Government Printing Office report, an author, Henry Hall, wrote," It is a remarkable fact that while Bath builds the cheapest wooden vessels in the United

States at present, nearly every ton of timber, iron, pitch, hemp, salt, canvas, etc., entering into their composition is imported into the town from places outside the state."

"Cheapest" in that context meant lowest in price not lowest quality. Low wages seemed to be the norm throughout Maine for several centuries. Workers managed to scrape by, the costs were kept down, and ships built there didn't cost as much as elsewhere.

The ship builders concentrated on cargo capacity to the detriment of sailing qualities. As coastal schooners grew from three masters up to four, five and six they were not required to maneuver in and out of harbors as there were numerous steam tugs to move them at all ports of every size. A schooner of one of the big coal fleets usually was reported by a maritime agent through the telegraph as she passed through Vineyard Sound or passed Highland Light and a tug was waiting off Boston or Portland the next day to tow her in. The business of the schooners was to move as fast as possible to make as many trips as they could each year and not to indulge in fancy sailing.

The pressure from the owners was to keep ships moving and thus make money. Many of the accidents to vessels sailing through Nantucket Shoals and Pollock Rip were caused by captains pushing vessels at times and circumstances when a safe anchoring would have avoided it.

In this photograph from the early 1900s these coal schooners wait in Portland Harbor for stevedores to unload earlier vessels, so they can take their turn at the docks. The big, five-master at the left in back is empty and ready for another trip south. Portland was the largest coal port in Maine.

The real growth in the coal trade Down East began in the 1870s. By 1894, a year for depressed business, 653,000 tons arrived in Portland. All the coal arrived by water, most of it in schooners. It was the cheapest way to ship anything. Portland was the state's leading manufacturing center and cheap coal was the power source. In New

England Boston was the leading coal port in tons received, Providence was next and Portland was third. In 1904 Boston received 4,100,000 tons of domestic coal by water. Coastal caravans of big steam tugs and schooner-barges, many of which were built in Bath by Kelley-Spear, outnumbered sailing vessels by two to one in carrying coal. Sails' great enemy was the schooner-barge. A big,steel coastal tug 160 feet in length could pull three barges, each with more than 4,000 tons. There was great economy in such an operation. The weather could be a factor in deliveries by schooners, but it rarely slowed down the tugs and barges. In fact, the steady movement of barge traffic caused a number of accidents to barges and a threat to their crews. In a twelve year period, beginning in 1890, 472 barges were lost on the Atlantic and Gulf coasts.

By 1900 nearly sixty steam tugs were operating in the port of Boston. The Reading Railroad of Pennsylvania was the largest supplier of anthracite coal to New England with fifteen big coastal tugs and more than one-hundred schooner-barges. The largest coal port was Philadelphia and both the Reading and Pennsylvania Railroad brought their coal there to the Delaware River piers. The other ports which received coal for New England were Baltimore, Newport News and New York. By 1906 more than 25,000,000 tons of coal entered New England by water every year. In fact, August Belmont, the builder of the Cape Cod Canal, decided to build his waterway primarily to capture a major share of that annual volume of coal heading north around Cape Cod.

The waters between Maine and the mid-Atlantic coal ports were extremely dangerous, with fog, shoals, strong tidal currents and rips, gales, hurricanes and blizzards. As a result there were numerous marine accidents which kept lifesavers busy.

After the Civil War the ice industry in Maine had its greatest growth following a shift in public taste from salted meats to fresh meat. On the banks of the Kennebec River above Bath was the center of the industry. The first attempts of building an ice industry date from 1831 when a Boston businessman, Frederick Tudor, tried to ship ice to the West Indies. Within twenty years the problems of cutting, storing and shipping the tidewater ice had been solved.

Ice that cost thirty cents a ton to harvest on the Kennebec in some years retailed for 50 cents a hundred-weight. Maine ice men built large vaults in Savannah and Charleston. After that they developed markets in Philadelphia and Baltimore.

There were huge icehouses, resembling giant barns along miles of the Kennebec River and the economic effect of the ice business was enormous. Building those icehouses required great quantities of lumber and carpenters to nail them together. The need for schooners to carry ice southward down the coast had its impact on shipbuilding and the shipping industry for years.

The year 1890 was a boom time in the ice business with 3,000 men working in it just around the Gardiner, Maine area. The following year there were forty-nine icehouses with a capacity of a million and a quarter tons along both banks of the Kennebec River north of Bath.

Some schooners carried as much as 1,000 tons of ice. It was shipped seven months a year. During that time the Kennebec River had more departures and arrivals then any other river in America.

For example, in 1887, 4,000 vessels, almost all of them schooners entered the river to load 930,000 tons of ice and 47 million board feet of lumber. Imports of coal totaled 115,000 tons. Many schooners came north with coal, New England's major import, and sailed south with ice, lumber or stone, a reference to granite, from quarries close to the ocean. Icehouses on rivers could not ship in the winter as the rivers froze over. Icehouses that stood between saltwater and freshwater ponds, as at Boothbay, Rockport and Sargentville, could.

These men are using horses to scrape an overnight snow off the ice in Dresden. Then the cutting will begin. This is the Cedar Grove ice house of the Cochran-Oler Company of Baltimore who built this huge building in 1885 to hold 70,000 tons.

The ice was poled to the apron of the elevator which carried the blocks to the top of the ice house for storage. This tidewater ice of the Kennebec River above Bath was the largest source in Maine, where the business thrived from 1870 to 1920.

In 1890 at Bank Street on the North River, Manhattan, a Maine-built four-masted schooner carried 2,500 tons of ice to be unloaded in New York. The shipment paid $1.50 a ton. The ship was built for $65,000. three and one-half years before. She eventually paid for herself more than twice over. By 1894 one schooner, the *Sarah W. Lawrence*, of Fall River, Massachusetts, owned in Taunton by a group there had paid a profit over her cost of $200,000. The natural ice industry was pretty much dead by 1910 in the seaboard cities having lost out to manufactured ice. A group of coastal shipping men in Taunton, Massachusetts between 1870 to 1895 had a fleet of handsome four-masters many of them built to their order and design in Maine shipyards. Men in Taunton owned forty three-masters and fourteen four-masters.

One of the good things economically about coastal shipping was the availability of freight going both ways, i.e., coal north, lumber south, coal north, ice south, coal north, granite south, coal north, lime south.

Burned lime, the chief component of plaster, was a construction material of great importance in growing cities.

The Kennebec River was also the avenue to the sea to move ice to market. Once the river thawed in April, shipments began south. The four-master, *Sarah Lawrence,* is loading ice in Dresden. Ships this size could carry as much as 2,700 tons. For a trip to New York the freight bill was $1.50 a ton. By 1900 more barges were used with their smaller crews being more economical. A coastal tug could pull three barges.

Rockland was the largest producer of lime for the construction industry with its major expansion after the Civil War. Lime was used in plaster and was shipped by water from these docks with schooners lined up for cargo from nearby kilns.

Typical "limers" were two-masted schooners who were bought cheap after long careers in other industries. Many were unsafe and near the end of years of sea-going. Quite a few burned.

Most lime came from Maine. Railroads brought the limestone from quarries nearby Rockland and Thomaston to the waterfront where lime kilns were located. The kilns were 28 to 36 feet high and were made of stone and lined with firebrick. Cordwood was lit under the limestone rock and wood was continually thrown onto the fire which burned the carbonic acid from the rock, leaving the lime, which was put in casks like small barrels and then shipped in schooners, called "limers."

Limers frequently caught fire. A sudden leak in a vessel and the cargo caught fire. Water and lime produced fire-and often the crew intentionally beached their ship. Such fires were hard to fight, as water could not be used-water only made things worse.

When there was a fire on a limer, the technique was to seal every crack with lime plaster and keep oxygen from the flames. If the fire died out, the trip continued; sometimes, if the vessel was far gone, she would be stripped of sails and other deck gear. If she was in danger of the fire breaking out, she would be scuttled.

Cobb, Wight & Co., Rockland's biggest lime manufacturer, lost seven lime schooners, six of them by fire.

The ideal lime coaster was old, cheap and light. Fishing schooners made good limers, because the shape of their hull placed the lime on a platform above the bilge water. In 1888 there were 275 vessels manned by 800 men involved in the Knox County, Maine lime trade. The ships were carrying 700 to 1,500 casks apiece. Lime produced that year was 1.8 million casks, suggesting more than one thousand cargoes. Smaller vessels were used to carry lime to Boston, but the bigger schooners were on the New York run, the largest market in America.

In 1889 the newspaper, Rockland Opinion, editorialized: "Rockland men are all the time buying and bringing down here a lot of old tubs from New York and Massachusetts when our shipyards are so well equipped to turn out new vessels of any class. Our sailors,too, ought not to go to sea in some of the old tubs that are used in the lime business; it is not safe, for they are liable to go to pieces in a little gale that would not trouble a good sound vessel at all." These old tubs often had accidents along the coast. Many of them were in a sorry condition with hulls that were leaky and limber, often requiring frequent pumping to remain afloat. When these older tubs grounded the swells caused bumping that opened up the hulls and the vessels sank. In fact these ships could start to fall apart in the heavy swells of a gale at sea. The editor had a good point.

In those days there were no inspections by government agencies, like the Coast Guard.

In 1898 there were 57 coasters sailing out of Rockland. Lime coasters were also owned at Rockport, Thomaston and Tenants Harbor.

Maine was also a leader in the stone business. The granite quarries around Hallowell of the Hallowell Granite Works had its cutting sheds near the Kennebec River for transport. Hallowell is below Augusta a few miles. The local soft, white granite was ideal for statues and the post-Civil War period was one of great sentimentality. Statues and monuments were a major part of their work and many wealthy people ordered granite mausoleums before they died. In 1888 the Hallowell Works employed 350 people carving granite. In 1894 pneumatic carving of granite was begun.

Carver's Harbor in Vinalhaven was a major shipping point with schooners being loaded there constantly. One-thousand men were employed on Vinalhaven in all phases of quarrying, moving, cutting and polishing granite.

The paving block industry started on Cape Ann, Massachusetts and spread quickly to Maine. Paving stones for roads were cut in small two-man operations called "motions." Most paving blocks were cut from waste stone. Maine paving blocks were shipped down the coast to Baltimore, Washington, Philadelphia. The biggest market was New York City. Western markets were Chicago and St. Louis with shipments by rail.

The early centers for cutting paving blocks were Vinalhaven, Tenants Harbor, Frankfort, Sullivan and West Sullivan. Some of the first granite blocks were a foot long, eight to ten inches wide and six inches deep. The widespread use of paving blocks began after the Civil War. The first cutters were Scottish, followed by English, Welsh, Irish, and finally Swedes and Finns.

Granite from local quarries was moved to the waterfront at Vinalhaven for cutting, polishing and shipping by water. Its designation as cargo was "stone."

The State of Maine produced more wooden ships for coastal transport than any other region of America. Those vessels started life carrying the products of their birthplace: lumber, ice, potatoes, lime, and granite.

Farther down the coast Massachusetts cargoes were sand from Plum Island, salt hay from Newbury, Ipswich and Newburyport, granite from Cape Ann harbors, fish from Gloucester, shoes and other manufactures from Boston, finished cotton goods from Fall River.

From the other direction came timber from the south and the largest north-bound cargo: coal, millions of tons of it from Newport News, Baltimore, Philadelphia and New York to feed New England's hungry furnaces. Most of this freight moved over Massachusetts water every day of the year, sustaining a coastal fleet of thousands of vessels.

Past lighthouses and lightships and watched over by surfmen walking lonely beach patrol through the darkness and men in towers at life-saving stations scanning the horizon the steady parade of ships sailed on into a history that is now but a memory.

Bangor shipped more lumber than any city in the northeast.

POINT ALLERTON

Point Allerton on the northern edge of the Hull Peninsula is a unique place in American lifesaving history. For so many years it was the scene of countless shipwrecks and spirited sea rescues. The Massachusetts Humane Society built two of its first three houses of refuge near there: one at Nantasket Beach and the other on Lovell's Island. The third was near the beach at Scituate. These small wooden huts were shelter and stocked with firewood and supplies for survivors of nearby wrecks. They were the beginning of a large lifesaving operation by that Society founded in Boston after the Revolutionary War.

The Humane Society had a boat house at Hull from at least 1840 and its surfboat, the *Nantasket*, was used in numerous rescues near there. In the six years after 1840 the *Nantasket* with its volunteer crews brought thirty-six survivors from wrecks to the shore. The most decorated lifesaver in America was a Hull native. Joshua James was born there in 1827 and as an adult became a volunteer lifesaver. The United States Life-Saving Service referred to him as our most decorated member, although James spent only the last thirteen years of his life as the keeper of the Point Allerton Station. Most of his career around the water was as a member of the Massachusetts Humane Society.

It was as keeper of the Hull Life-Saving Station of the Humane Society that Joshua James set an outstanding record of sustained rescues when he noticed on the afternoon of November 25, 1888 that a number of sailing vessels anchored off Nantasket were dragging their anchors. This was the beginning of a violent northeast storm that pounded ships at sea and the coastline. To the north of Hull is Toddy Rocks and it was there the schooner *Cox And Green* stranded and was in danger of breaking up in the violent surf. James called out his volunteers and they fired a line with their Hunt Gun and followed with a breeches buoy. All in the crew of the *Cox and Green* were brought ashore to safety.

While this rescue was taking place, a short distance to the east another schooner was on the rocks and flying a signal of distress in her rigging. She was the *Gertrude Abbott* with a cargo of coal for Boston. She was too far off shore for a breeches buoy rescue, so James and his volunteers would have to launch a surfboat. Joshua James after looking at those mountainous seas warned his men they might not live through the attempt. Each man in his volunteer crew stepped forward.

After an exhausting pull and with two men bailing constantly, the lifesavers got close enough to toss a line to the sailors who slid down ropes into the surfboat. The return to the shore was more hazardous as the rescue craft struck rocks which stove holes which threatened to sink her. The surfboat was pushed along by the heavy waves and finally struck more rocks in the shallows. The eight sailors and James and his crew jumped out and started wading for the shore, which all of them made safely.

In the middle of the night a surfman on beach patrol saw another schooner in a bad place. It was the *Bertha F. Walker* whose crew of seven had climbed into the

The surfboat *Nantasket* with eight men at the oars and Joshua James at the steering oar is a painting by John M. Kilroy. This craft, designed by a brother of the keeper, was built in 1887 and is on display in the boatroom at the Hull Life-Saving Museum. This picture depicts the technique of ramming a surfboat straight ahead through high waves. If it didn't work, it would be tried again until it did.

The schooner *H. C. Higginson* stranded near Nantasket Beach on November 25, 1888. Joshua James and his crew from Point Allerton rowed to the vessel through heavy seas and rescued five men who were in the rigging.

This is a popular and widely-published photograph of Joshua James, the renowned keeper of Point Allerton Life-Saving. His phenomenal record may have reached eight-hundred lives saved. He had more medals for his deeds than a Russian General after Stalingrad. The U. S. Life-Saving Service regarded him as their star, yet James spent most of his career earlier as a volunteer with the Massachusetts Humane Society. His was a life of dedication.

Below: This is the original Life-Saving Station at Point Allerton built in 1889, facing the south channel entrance to Boston Harbor. It was an important station, and was close to the scene of so many shipwrecks. It is now the Hull Life-Saving Museum.

Pt. Allerton U.S. Life Saving Station
Hull, Massachusetts, c. 1905

The *Henry R. Tilton* on Toddy Rocks, Hull, was testimony to the power of the *Portland* Hurricane. The *Tilton* was blown ashore one and a half miles west of the Point Allerton Station. Joshua James and his surfmen pulled the beach apparatus overland to rescue the crew of seven men with a breeches buoy, and then took them to the station for shelter.

Below: The Humane Society's crew at Point Allerton with their surfboat, *Nantasket*, designed by Samuel James, an older brother of Joshua. At the right is the standard issue beach cart with a Hunt Gun on top.

rigging to escape the waves crashing over her deck. With their surfboat wrecked, James and his men secured some horses and had to pull a surfboat four miles over land. After dawn Joshua James and his crew, including a few new men, rowed to the *Walker* and brought all seven men to shore.

When James and his volunteers were landing the crew of the *Walker*, a horseman rode up with the news of two more strandings. The nearest one was the schooner *H. C. Higginson.* The Humane Society crew reached a spot opposite the wrecked vessel the same time a crew of surfmen from the United States Life-Saving Service Station at North Scituate arrived. The latter group had pulled their beach apparatus nine miles through the snow. James and his volunteers had a Hunt Gun with their apparatus and a surfboat.

Both groups of lifesavers fired their guns simultaneously toward the *Higginson.* Her sailors grabbed the Humane Society's line and tied it to one of the masts. When the line soon fouled, a rescue by breeches buoy was ruled out, so Joshua James and his men launched their surfboat which suffered severe damage on rocks and had to be repaired. A second try was attempted through heavy seas and bitter cold. This time they finally reached the *Higginson* and one by one the sailors dropped into the surfboat. Out on the water it was a slow rescue, as the five survivors were exhausted in that brutal cold and high wind. Once on the beach, the Society volunteers brought the five back to shelter and warmth.

On their way back to the boathouse, they came upon an abandoned sailing vessel breaking up near the shore. There were several trapped men on board trying to salvage the ship. Keeper James noticed their plight and rescued the pair. He and his men had taken part in reaching five shipwrecks and bringing twenty-nine sailors to safety.

His organization, the Humane Society, and the U.S. Life-Saving Service each presented James with a gold Medal. He would earn five more gold and silver as America's most-honored lifesaver.

The following year the Life-Saving Service opened its Point Allerton station and chose Joshua James to be keeper, even though at age sixty-two he was years past the entry age for keeper. He knew the area and had become a legend around Hull, Nantasket, Boston Harbor-and beyond.

For the next thirteen years Joshua James, the quintessential surfman, was the keeper of the busy Point Allerton station. He rarely left the station and maintained his dedication at dozens of maritime emergencies off those shores. The wildest test for him and the other Massachusetts lifesavers came during the three days of November 26-28, 1898, the *Portland* Storm."Every man that was alive at the time we started for him" was successfully rescued during that frightening violent weekend. Those words from Joshua James in his report to the Life-Saving Service. Details of their work is included in the chapter on that storm.

It was four years later that James died on the beach near his station at Point Allerton. He had called out his men for a scheduled surfboat drill and he had insisted it be unusually vigorous. James was distressed by the loss of seven lifesavers from the Monomoy station at the stranded coal barge *Wadena* off Cape Cod just two days before. That tragedy was one of the worst in the history of the Service. It bothered him greatly.

After the drill Joshua James put down his sweep oar and stepped from the surfboat onto the beach. After taking a few steps he turned toward his crew and reportedly spoke four words: "The tide is ebbing." He fell dead. He was seventy-four.

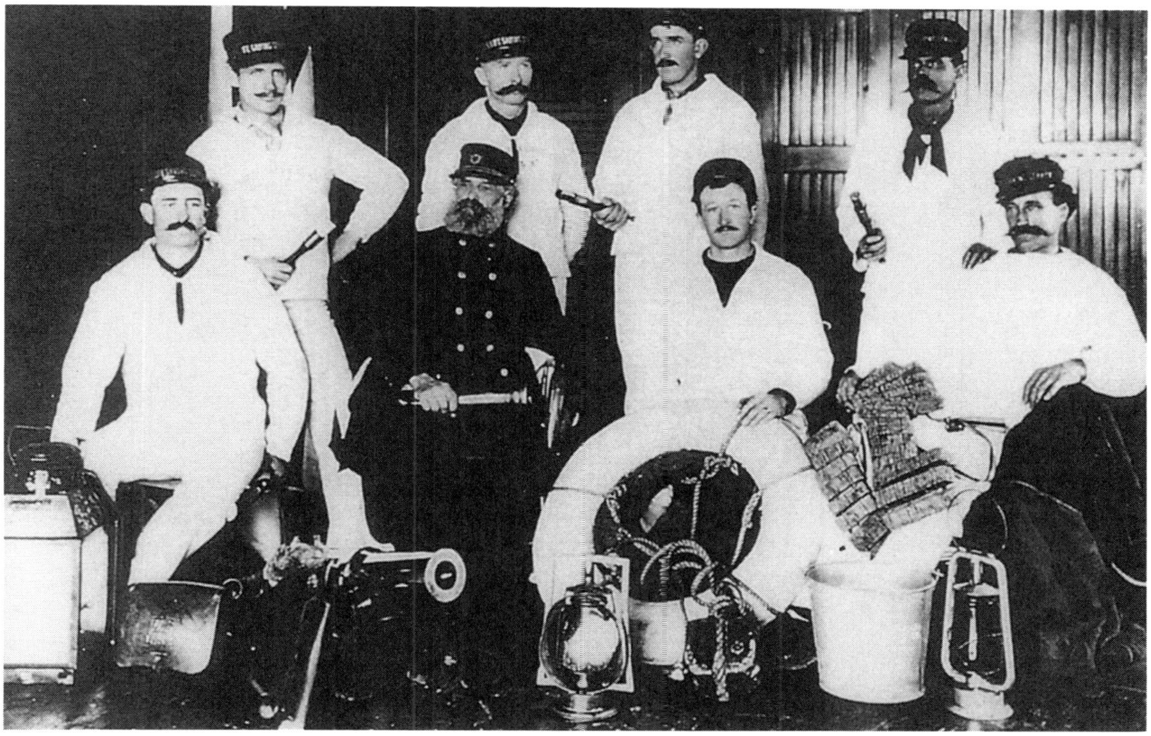

This is a typical posed photograph at many stations. The keeper, Joshua James, in his black uniform; the surfmen in their summer whites with those in back holding Coston flares; two of the surfmen on the right front with a life ring and a cork life jacket. The other equipment from the left is a beach lantern, a Lyle gun, and several portable barn lanterns. The keeper is holding a speaking trumpet for communication during noisy rescues.

THE *NANCY* AT NANTASKET

Oh, those northeast storms, most of them gales, always raised havoc with ships and the Massachusetts shoreline. A violent gale with winds of seventy miles an hour damaged and sank a number of vessels caught at sea on February 21, 1927. Several Coast Guard cutters were lost around Cape Cod.

The most famous shipwreck from that storm was the stranding of the five-masted coal schooner, *Nancy,* on Nantasket Beach. The big schooner-she was 259 feet long and had a gross weight of 2,117 tons-had unloaded her cargo of Virginia coal and was heading light back to Norfolk for more. Her captain should have kept her anchored in Boston Harbor, but he decided to sail too late. With the barometer dropping in the face of heavy and rising winds, two anchors were dropped to the east outside *Boston Lightship*.

The storm created mountainous waves that pounded the high sides of the *Nancy* with all that windage and she dragged across open water and grounded in the surf, a big spectacular vessel right on the beach where everyone could see her. And thousands came to see her.

But not before lifesavers showed up to rescue her crew of nine. The Coast Guard was called, but they never showed. So a volunteer crew of the Massachusetts Humane Society under Captain Ose James reacted in the best tradition of Hull and its Point Allerton Station. Witnesses thought it was a scene from the past with borrowed horses, snorting steam, pulling a carriage with a surfboat along local streets to a point opposite the *Nancy* where they could launch.

There was a call along the beach for one more man. Robert Blossom, a firemen from Hull, volunteered and after taking his seat along the thwart wondered if he had made the right decision. He later described himself as "young and foolish."

Picture this, if you will. The visibility was poor. The winds were close to hurricane velocity. There was at least one foot of snow on the ground and the waves were mountainous. Into all that turmoil the nine Society volunteers rowed out to the lee side of the schooner, which was broadside to the beach. The waves crashed against the exposed port side and rolled across the deck and onto the surfboat below.

The crew of the schooner were throwing over duffel bags and suitcases with their belongings. The men in the surfboat were picking them out of the water with their hands or with boat hooks.

There was a rope over the starboard side of the *Nancy* down which her crew slid one at a time. The surfboat was rising and falling and on the rising wave the men waited then jumped for it. The next to last man to leave the schooner was quite old and very frightened. The volunteers later said he was the cook. After several opportunities to jump passed and the man still did not, a volunteer had to pry his hands off the rope and take him into the rescue craft. The captain was the last man off and then the nine volunteers pulled for the shore. There was a rope from the beach to the surfboat, but the incoming waves were so high that the boat was brought onto the beach on a retreating, or receding, wave, which was much safer rather than riding a comber in. The volunteers waited for a wave to break and run up the beach, then they pulled quickly to the sand and jumped out where the many willing hands pulled the surfboat by the rope and helped the eighteen men in the boat to jump out to safety.

Robert Blossom said, "We were lucky, because the next wave smashed the surfboat against the rocks and if we had been in it, we could have drowned. The water was that rough."

Ralph Rich, the captain of the local Coast Guard, said afterward, "It's lucky that you guys went out, 'cause half my men can't row. They don't know how." The Humane Society volunteers knew and for them it was an emergency to be answered.

Blossom had a 1923 Ford. His clothes were so frozen to him by the spray that his feet could not move the pedals, so he asked someone to drive him home. It was that bitterly cold.

The nine year old schooner, built in Portland, Oregon with a home port in Philadelphia was owned by a New Jersey woman, Miss Clema Collins.

A gangway was built from the beach to the deck on her port side and thousands paid twenty-five cents each to walk up and go aboard the famous wreck; children paid ten cents less. To see the *Nancy* was another reason to visit the popular Nantasket Beach.

The Coast Guard was called, but never showed, so a blue-ribbon crew from the Massachusetts Humane Society at Hull rowed its surfboat out to the coal schooner *Nancy*, stranded at Nantasket Beach, February 21, 1927

The volunteers brought the crew of eight men to safety on the beach. The men had thrown overboard their suitcases, which were recovered.

The *Nancy* became a most popular tourist attraction with a ramp to bring visitors on board. Adults twenty-five cents, children fifteen. A photograph of you and the *Nancy* was a dollar. After many years the masts came down and she was finally burned.

BOSTON HARBOR PORTFOLIO

Sailing vessels needed a tug to help them enter and leave Boston Harbor and most other ports. The schooner *Albert H. Willis,* loaded with Maine lumber, is assisted by the tug *Powow* of the Ross Towboat Company of Boston. The *Willis* had a crew of eight; the *Powow* had four: two below and a captain in the pilot house and a deck hand to handle lines. R. Loren Graham of Swampscott, Massachusetts roamed all over Boston and other ports with a camera and his trained eye for a good picture.

Revenue cutters had maintained patrol of the Massachusetts coast since the service was founded by George Washington. They were originally intended to prevent piracy and enforce import duties. They often helped the lifesavers when they were in the vicinity. This cutter, of the Manning class, was a familiar vessel along the Massachusetts coast.

Up to America's entry into World War II some coal was still brought to Boston occasionally by sail. The *Mary Manson Gruener* had a tug on her port side as she is being moved to a Chelsea dock.

This is the floating lifesaving station at City Point, Dorchester Bay, Boston that was placed there in 1896 by the U. S. Life-Saving Service , because of the large number of fatal accidents to pleasure craft during warm weather. The nine surfmen there were constantly busy, especially weekends, chasing wayward craft and boats with the new gasoline engines that often failed. Many of the amateur yachtsmen in peril from tides, currents and sand bars knew little basic information about the water and small boathandling. This station had one of the first gasoline-powered surfboats the Service purchased. The replacement for this station lasted until 1950.

Boston had lots of steamboats up to World War II. In the foreground is the *City of Lynn*, a ferry crossing the harbor with commuters to take the Boston, Revere Beach & Lynn Railroad, a narrow gauge line that everyone called "The Gauge." The white paddle-wheeler is the *Nantasket*, one of many excursion boats that sailed the harbor day and night is pulling away from Rowe's Wharf. That single stack at the left is from one of the Metropolitan Steamship Company's India Wharf New York-bound boats, the *H. M. Whitney* or one of her sisters.

This is Little Brewster Island across the south Channel from Point Allerton, Hull. There has been a beacon there since 1679. The British blew it up during the Revolution. It was here that a cannon was fired during heavy fog as the first such warning in that weather. It is America's oldest lighthouse location.

Boston was New England's largest coal port with millions of tons arriving annually from as far south as Newport News, Virginia. These two five-masters, the *William C. Carnegie, center*, and the *Jennie French Potter*, are waiting their turn to unload. Other wind ships arrived later and will wait longer, as the rule at both ends of the trip was "absolute turn."

OTHER SEA RESCUES

These crewmen from the schooner-barge, *Ashland,* were rescued at North Scituate by the Coast Guard during a savage northeast storm March 4, 1916 after their vessel and another barge, the *Kohinoor*, parted their lines from a tug and broke up on ledges. The rescuers fired a line to this deckhouse and brought five of them to safety. The four men on the *Kohinoor* did not survive.

The Boston-New York steamship *Robert E. Lee* was running toward the Cape Cod Canal, March 9, 1928, and wandered off course in a blinding snowstorm and struck Mary Ann Rocks off Manomet Point. She was rolling in heavy swells until the seacocks were opened and she settled. The next morning a surfboat from the Manomet Station rowed out to the *Lee*. It was considered too dangerous to try to ferry 273 passengers to the station, so power lifeboats were brought from the Sagamore and Wood End Stations on the Cape. These were used to move all passengers to three patrol boats that had arrived. When the surfmen from Manomet rowed back to their station, a huge wave struck and another, capsizing the small craft. Three surfmen drowned; the others survived.

The Coast Guardsmem from the station at Old Harbor, Cape Cod, are bringing the crew of the grounded Canadian freighter, *Lutzen,* in a surfboat to North Beach. This is six months before World War II, but it looks like an old time rescue with men wearing life preservers and rowing toward the beach. The *Lutzen* carried more than 200-tons of frozen blueberries, some of the cargo was salvaged.

The crew of the *Lutzen* tried to come ashore in this wooden lifeboat, which didn't make it. However, the Coast Guard rescued everybody, and then posed beside the broken lifeboat.

The Panamanian-registered freighter, *Evgenia*, grounded on shoals at Peaked Hill Bars, Provincetown during a storm early in September, 1953. The Coast Guard responded with a dramatic rescue.

This Greek sailor from the *Evgenia* is the twelfth man brought ashore by breeches buoy by the Coast Guard who responded from Provincetown to Chatham. The line snapped when the next man was heading for shore. He fell into the surf and was rescued. Thirteen men were saved this way; four were brought ashore in an amphibious vehicle and seven stayed aboard the ship. The breeches buoy was used but once more on the Cape, then retired.

In the winter of 1931 the original Peaked Hill Bars Life-Saving Station fell over the eroded bank and washed into the sea. Its replacement is down the shore in the background. The building that tumbled here was once the home of Eugene O'Neill where he wrote some of his earlier plays. One of those plays was first produced in a makeshift theater on a wharf in Provincetown.

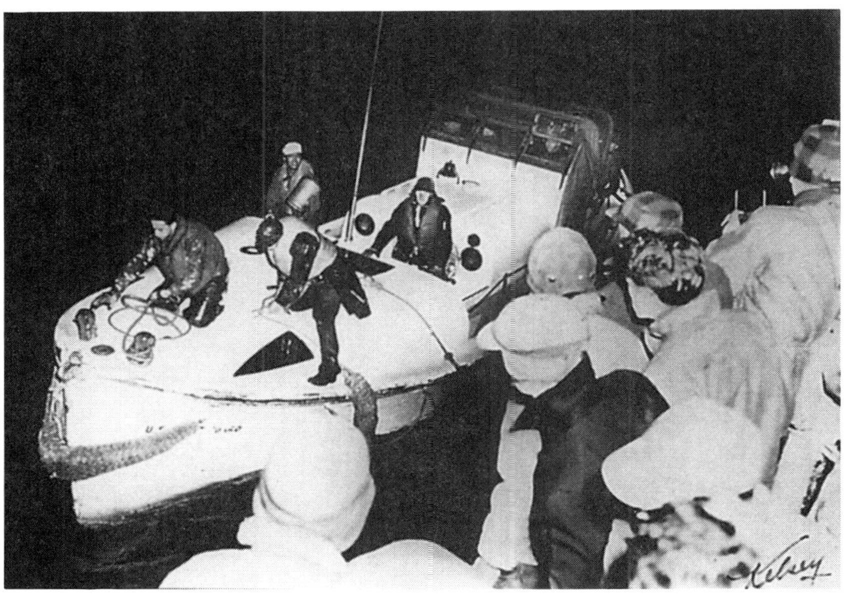

An incredibility violent storm in the Atlantic broke the tanker, *Pendleton*, in half a few miles off Cape Cod, February 18, 1952. A Coast Guard crew of four from Chatham led by Boatswains Mate Bernard C. Webber left the Chatham Station in a 36-foot motor lifeboat. After a hazardous trip through 60-foot high waves, the rescue craft circled the stern of the tanker where 33 men were and came along side. All but one man climbed down a ladder safely and into the lifeboat. The other man fell and was lost.

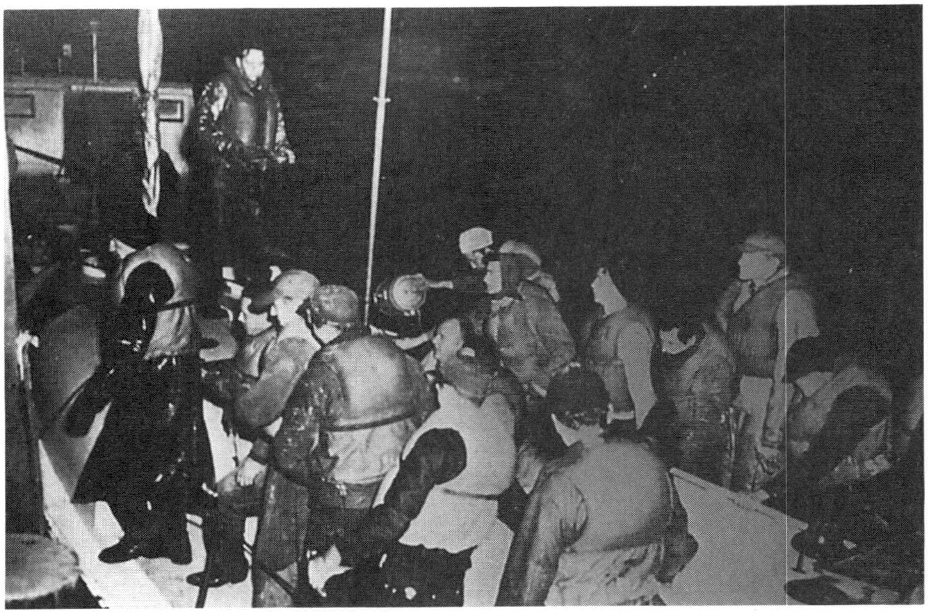

At the pier in Chatham the survivors leave the craft that rescued them.

The next day a Coast Guard cutter saw the bow of the *Pendleton*. Some miles off shore the tanker *Fort Mercer* also broke in two in the same storm. A Coast Guard cutter rescued some of her crew, but five men were lost.

NANTUCKET

The first lifesaving station on Nantucket was built in 1874 at Surfside on the ocean south of the Town of Nantucket. Many other Massachusetts stations were erected with this same design. The lookout was not enclosed and was reached by a ladder inside the second floor. A tour of duty as lookout in the winter could be a grim experience.

In 1884 this station was built at Coskata, Nantucket.

In the cupola on watch is surfman Erastus Chapel of the Muskeget Life-Saving Station. It was an enclosed room, but he still had to use a ladder to come and go.

This is an unusual photograph because it shows the Francis Lifecar carrying a surfman back to the beach during a weekly drill. The Lyle Gun is between the faking box and the beach cart.

One-room sheds like this had a key for a time-clock that surfmen carried on their nightly patrols during the ten months of the active storm season, August 1 to May 31. The four lifesaving stations on Nantucket were too far apart to split the distance for the patrol with a surfman from an adjoining station to meet and exchange brass checks, as was the usual practice where stations were five miles apart, as in many Massachusetts locations.

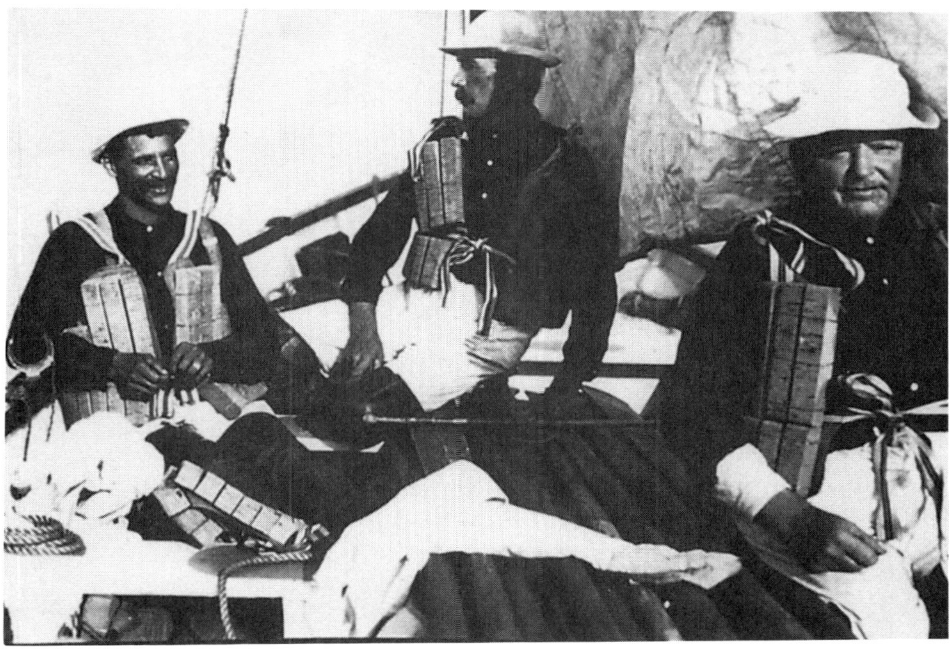

Three of the surfmen from the Muskeget Station during a drill with the sail hoisted in the surfboat. On the right is Erastus Chapel. In the center is Arthur Dunham. The man at the left is identified as a Mr. Abbott, who may have been the only black man in the U. S. Life-Saving Service on Nantucket around 1910.

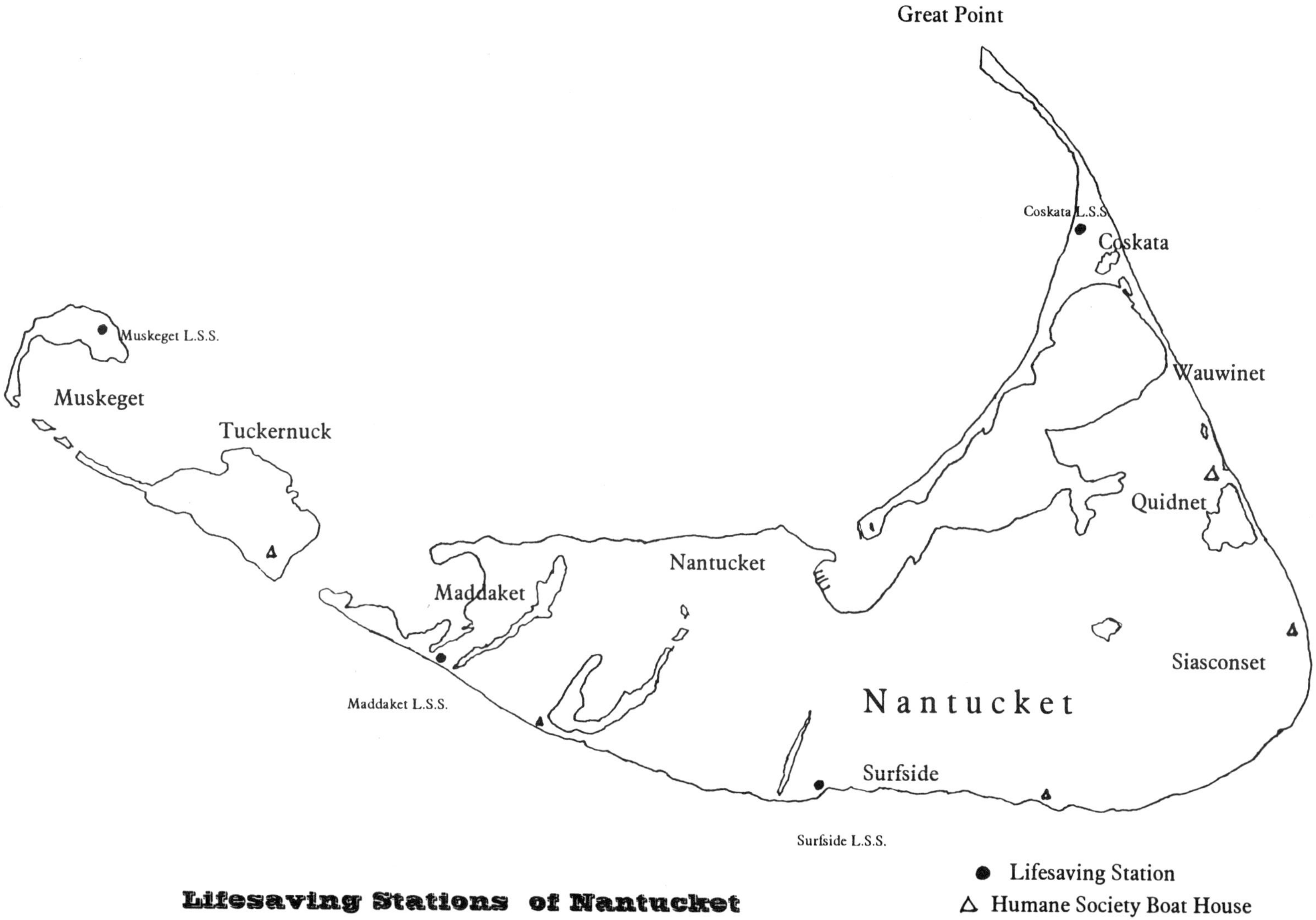

Lifesaving Stations of Nantucket and Humane Society Boat Houses

Nantucket had its share of foggy days and on March 9, 1877 this vessel, the bark *W. F. Marshall*, lost its way in a heavy fog and grounded on the south side of Nantucket. The surfmen brought all sixteen people ashore in the breeches buoy. The *Marshall* broke up in heavy surf that pounded her for days.

That big coastal tug, upper right, is out there, because somebody's in trouble and surfmen are launching a Jersey sea-skiff as a surfboat. The last surfman climbs over the stern and on the left the keeper will give a final push and be the last man aboard. Then hard rowing will get them out there, too.

This Nantucket surfboat is landing. The keeper with the long, steering oar is in the stern. It took great strength to handle a small boat this way.

Above: This is one of the larger Nantucket surfboats landing in big surf, always a dangerous time. The men on the right have been rescued from a wreck and are the first to jump onto the beach where they turn to help bring the boat out of the water to safety farther up on the sand.

Left: There were so many wrecks on Nantucket. This is the schooner *Oregon* at Wauwinet on a late summer day in August, 1885. She had a hold filled with Cape Ann paving stones. The granite was recovered and used on a street in Nantucket Town. A few of her timbers found new duty in a summer house near the wreck scene.

This is a supply trip for the Life-Saving Station on Muskeget Island. The motor yacht, *Sadie*, has come from Nantucket Harbor, bringing coal in sacks to heat the station and for cooking. The coal is in those dark bags on the beach to the right. Other supplies, including food and mail, came to the isolated station on the *Sadie*, which had a hoist rigged to that mast for lifting.

Here is a perfect example of a Francis Lifecar, made of metal with an open space inside for rescued people to lie down. The rope to and from a ship like a breeches buoy line ran through those two metal loops on top. There were rings, one on each end, to which a line was attached to pull the lifecar to a shipwreck and back to the beach. The lifecar was standard issue to every station. Their range was not more than six-hundred yards, however, they were rarely used. Surfmen had to drill, using lifecars, but this apparatus was not a favorite device. This lifecar is on display at the Nantucket Life Saving Museum. The Francis Company also made metal surfboats.

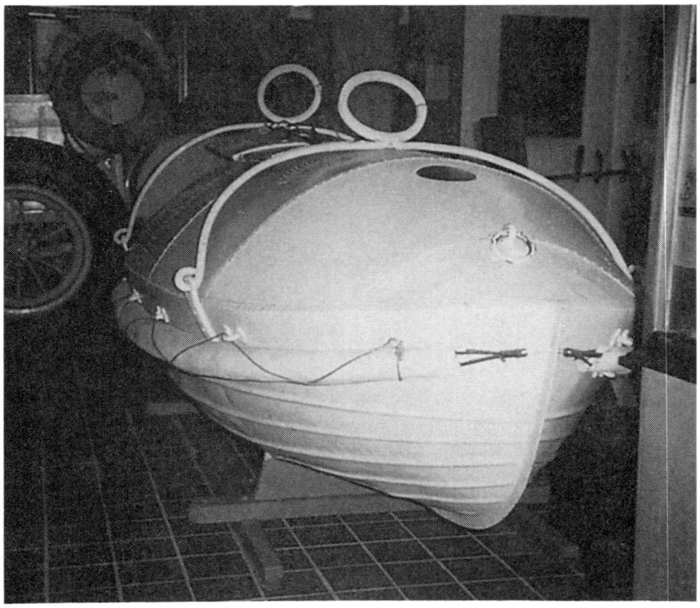

Maurice E. Gibbs, the President and Director of the Nantucket Life-Saving Museum, holds a cutting block which was used to sever the line after a rescue by breeches buoy or a lifecar. Gibbs is President of the U. S. Life-Saving Service Heritage Association, a national group with headquarters in Caledonia, Michigan. His museum on Nantucket is a fascinating collection of the treasures of American lifesaving, including a Race Point design surfboat, a type used at Surfside and Coskata, beautifully restored by Peter Culler. This small craft looks as if a master boatbuilder completed it ten minutes before you arrived.

The Nantucket Life-Saving Museum, opened in June, 1972, is a replica of the Surfside Station, but without the second floor. The plans were obtained from the Library of Congress.

THE SAGA OF HAROLD FLAGG

Harold W. Flagg left his job on Standard Oil Company tankers to join the Coast Guard in World War II. It was July, 1942. After boot camp, Flagg was promoted to Chief Boatswains Mate and that October was assigned to duty aboard *Lightship #73, Vineyard*, on station at the entrance to Vineyard Sound about two miles off Cuttyhunk Island. The lightship had a complement of seventeen men.

There was a regular rotation on leave of five men at a time. Harold Flagg was on leave with four others all of whom were to report back to the ship at noon on Friday, September 15, 1944. During the night before and early that morning a savage Atlantic hurricane struck. Sometime between 1 and 3 a.m. the *Vineyard Lightship* vanished off the Army's radar on Cuttyhunk Island. Not far away in Buzzards Bay the *Hen and Chickens Lightship*, #86, dragged anchors and was blown more than a mile off her station. Crew members on #86 later reported they had seen red and white flares from the direction of the *Vineyard Lightship*.

Harold Flagg took a train from Boston that Friday morning that went right out on the pier at New Bedford. He and the other four crew members were told: "We can't find the ship. Go home until tomorrow."

The following day they went out on small Coast Guard cutters to look for survivors. There were none. One body was picked up Saturday; another on Sunday and three more in the next few days. The other seven were never found. It was grim work, looking for people you knew.

Flagg and the other four men were transferred to Boston where they received all new gear to replace what was lost on the ship. In November, he was assigned to the Monomoy Point Life-Saving Station where he stayed for one year. They did daily drill with the lifesaving gear and walked beach patrol every night. Until V-E Day, we were worried about saboteurs landing from German submarines. In the late fall of 1945 he was assigned to the East Orleans Station where he was in charge of caretaker status, as the station was closed. He and two others each received $.65 a day Commissary Allowance. It was hard to eat well on $1.95 a day.

They were there at East Orleans until February. It was boring just hanging around, so they took Springfield 30-ought-6 rifles from the armory with plenty of ammunition and went out to the dunes and shot tin cans and owls sitting in the sand. "We weren't trying to kill the owls, just scare them," Flagg says.

The *Vineyard Lightship, #73*, was anchored at the entrance to Vineyard Sound two miles west of Cuttyhunk Island and four miles from Gay Head, Martha's Vineyard. She was 129-feet long, had an iron hull, a steam engine and was built in 1901 in Baltimore, Maryland.

On a warm spring day in 1943 a few crew members pose for this picture by the smoke stack of *Vineyard Lightship*. Harold Flagg is in the back, without a cap.

This is the Coast Guard Station at Monomoy Point in the spring of 1945.

After a weekly drill with the beach apparatus, four surfmen assume a pose for the camera. Harold Flagg is on the left.

It is winter of 1944-45 and beach patrol was a daily exercise from dusk until dawn. These men, armed with bolt action, Springfield rifles, and live ammunition have one job: to look for German saboteurs landing from submarines which had happened on the coast of Maine, Long Island and Florida.

This is Jiggs, the mascot of the Monomoy Point Coast Guard Station. Early in 1945 Jiggs was given to the station by a Chatham woman who said he might brighten up things at that lonesome outpost. He did.

At a lighthouse symposium in Hyannis in 1996, Harold Flagg, center, is flanked by Ken Black, on the left, Director of the Shore Village Museum, Maine's Lighthouse Museum in Rockland, Maine, and Richard T. Walker, a civilian engineer with the Coast Guard, who recovered the bell, shown here, from the *Vineyard Lightship* seventy-feet down in Buzzards Bay.

When Harold Flagg showed me these photographs and told the story you have just read, he said at the end: "Not a day goes by that I don't think about what happened to that lightship, what happened to my shipmates who were lost, and what could have happened to me."

The Sea is Woman, the Sea

is Wonder-/ Her other

name is Fate !

Edwin Markham

Index- People

Index-Vessels

Index- Organizations

BIBLIOGRAPHY

Albion, Robert G., Baker, William A. and Labaree, Benjamin W. *New England and the Sea*. Middletown, CT: Published for Mystic Seaport, Inc. by Wesleyan University Press, 1972.

Annual Reports of the U. S. Life-Saving Service, 1875-1915. Washington, D. C.: Government Printing Offfice.

Bachelder, Peter Dow and Smith, Mason Philip. *Four Short Blasts*. . Portland, Me.: Provincial Press, 1998.

Baker, William A. *History of the Boston Marine Society, 1742-1967*. Boston: Boston Marine Society, 1968.

Barnett, J. P. *The Lifesaving Guns of David Lyle*. Plymouth,In: Town and Country, 1976.

Bearse, Sr., Clarkson P. *The Tragedy of Monomoy Beach*. Harwich: The Goss Print, 1943.

Bergan, William M. *Old Nantasket*. Hull, Massachusetts: Hull Life-Saving Museum, 1968.

Beston, Henry. *The Outermost House*. New York: The Viking Press, 1928.

Bosworth, Janet (Editor). *Cuttyhunk and the Elizabeth Islands*. Cuttyhunk Island: Cuttyhunk Historical Society, 1993.

Bunting, W. H. *A Day's Work, Part I. 1860-1920*. Gardiner, Me: Tilbury House, 1997.

________ *Portrait of a Port:Boston, 1852-1914*. Cambridge: Harvard University Press, 1971.

Cahoon, Archelus H. and Paine, Mildred B. *Harwich Men of the Sea*. Harwich: Harwich Historical Commission, 1977.

Claflin, James. *Lighthouses and Lifesaving Along the Massachusetts Coast*. Charleston,S.C.: Arcadia Publishing, 1998.

Clark, Admont G. *Lighthouses of Cape Cod, Martha's Vineyard and Nantucket*. East Orleans: Parnassus Imprints, 1992.

Dalton, John W. *Along The Coast*. Boston: 1909.

______*The Lifesavers of Cape Cod*. Boston: Barta Press, 1902.

Everson, Jennie G., *Tidewater Ice of the Kennebec River*. Freeport, Me.: Bond Wheelright Co., 1970.

Fairburn, William Armstrong. *Merchant Sail*. Center Lovell,Me.: Fairburn Marine Educational Foundation, 1945.

Farson, Robert H. *Cape Cod Canal*. Second Edition. Yarmouth Port: Cape Cod Historical Publications, 1993.

_______*Cape Cod Railroads Including Martha's Vineyard and Nantucket*. Yarmouth Port: Cape Cod Historical Publications, 1993.

Fish, John Perry. *Unfinished Voyages, A Chronology of Shipwrecks*. Orleans: Lower Cape Publishing Co., 1989.

Freitas, Fred and Ball, Dave. *Warnings Ignored. The Story of the Portland Gale. November, 1898*. Freitas and Ball, Publishers, 1995.
Humane Society of Massachusetts. Report 1892.
Hough, Jr., George A. *Disaster on Devil's Bridge*. Mystic, Ct.: The Marine Historical Association, 1963.
Howe, M. A. DeWolfe. *The Humane Society of the Commonwealth of Massachusetts. An Historical Review.1785-1916*. Boston: The Riverside Press, 1918.
Junger, Sebastian. *The Perfect Storm*. New York: W. W. Norton & Company, 1997.
Kittredge, Henry C., *Cape Cod, Its People and Their History*. Boston: Houghton Mifflin Co., 1930.
Kirkland, Edward Chase. *Men, Cities and Transportation*. Two Volumes. Cambridge: Harvard University Press, 1948.
Lombard, Jr., Asa Cobb Paine. *East of Cape Cod*. Cuttyhunk Island: ,1976.
Merchant Vessels of the United States. 1870-1940. Bureau of Navagation, Treasury Department. Washington, D.C. Government Printing Office.
Merryman, J. H. *The United States Life-Saving Service-1880*. Golden, Co.: Outbooks, 1981.
Morison, Samuel Eliot. *Maritime History of Massachusetts. 1783-1860*. Boston: Houghton, Mifflin Co., 1921.
Morris, Paul C. *Maritime Nantucket*. Orleans: Lower Cape Publishing Co., 1996.
______*Schooners and Schooner-Barges*. Orleans: Lower Cape Publishing Co., 1984.
Noble, Dennis L. *That Others Might Live*. Annapolis: Naval Institute Press, 1994.
Quinn, William P. *Shipwrecks Around Cape Cod*. Farmington, Me.: Knowlton McLeary Co., 1973.
______*Shipwrecks Around New England*. Orleans: Lower Cape Publishing Co., 1978.
Regulations for the Government of the Life-Saving Service of the United States, 1899. Washington, D.C.: Government Printing Office, 1899.
Rockport As It Was. A Pictorial History of Rockport, Massachusetts: Sandy Bay Historical Society, 1975.
Shanks, Ralph and York Wick. *The U.S. Life-Saving Service*. Petaluma, Ca.: Costano Books, 1996.
Small, Isaac M. *Shipwrecks on Cape Cod*. Chatham: Chatham Press, 1928.
Smith, Jr., Fitz-Henry. *Storms and Shipwrecks in Boston Bay and the Record of the Lifesavers of Hull*. A paper read before the Bostonian Society. December 19, 1916.
Snow, Edward Rowe. *New England Sea Tragedies*. New York: Dodd, Mead & Co., 1960.
______*Storms and Shipwrecks of New England*. Boston: Yankee Publishing Co., 1943.
Stackpole, Edouard A., *Life Saving Nantucket*. Nantucket: Nantucket Life-Saving Museum, 1976.
Sullivan, Robert F. *Shipwrecks and Nautical Lore of Boston Harbor*. Chester, Ct.: The Globe Pequot Press, 1990.
Thoreau, Henry David. *Cape Cod*. New York: Penguin Books, 1987.
Van Dorp, Will and Lee, Linda. *Incomplete Journeys On and Off Plum Island*, 1873-1938. Newburyport: Rowley Press, 1989.

Weare, Nancy V. *Plum Island The Way It Was*. Newbury: Newburyport Press, 1993.

Wood, Vincent L. *Plum Island Recollections, Circa 1900*. Newbury: Newburyport Press, 1995.

Wreck and Rescue. The Journal of the U. S. Life-Saving Service Heritage Association. Caledonia, Michigan.